AF158252

Tucholsky Wagner Zola Scott Sydow Freud Schlegel
Turgenev Wallace Fonatne
Twain Walther von der Vogelweide Fouqué Friedrich II. von Preußen
Weber Freiligrath Frey
Fechner Fichte Weiße Rose von Fallersleben Kant Ernst Frommel
Richthofen
Engels Fielding Hölderlin
Fehrs Faber Flaubert Eichendorff Tacitus Dumas
Eliasberg Ebner Eschenbach
Feuerbach Maximilian I. von Habsburg Fock Zweig
Ewald Eliot Vergil
Goethe London
Mendelssohn Balzac Shakespeare Elisabeth von Österreich
Lichtenberg Rathenau Dostojewski Ganghofer
Trackl Stevenson Doyle Gjellerup
Mommsen Tolstoi Hambruch
Thoma Lenz Hanrieder Droste-Hülshoff
Dach Verne von Arnim Hägele Hauff Humboldt
Karrillon Reuter Rousseau Hagen Hauptmann Gautier
Garschin
Damaschke Defoe Hebbel Baudelaire
Descartes Hegel Kussmaul Herder
Wolfram von Eschenbach Dickens Schopenhauer Rilke George
Darwin Melville Grimm Jerome Bebel
Bronner Campe Horváth Aristoteles Proust
Bismarck Vigny Voltaire Federer Herodot
Gengenbach Barlach Heine
Storm Casanova Tersteegen Gilm Grillparzer Georgy
Brentano Chamberlain Lessing Langbein Gryphius
Strachwitz Claudius Schiller Lafontaine Iffland Sokrates
Katharina II. von Rußland Bellamy Schilling Kralik
Gerstäcker Raabe Gibbon Tschechow
Löns Hesse Hoffmann Gogol Wilde Gleim Vulpius
Luther Heym Hofmannsthal Klee Hölty Morgenstern Goedicke
Roth Heyse Klopstock Puschkin Homer Kleist
Luxemburg La Roche Horaz Mörike Musil
Machiavelli Kierkegaard Kraft Kraus
Navarra Aurel Musset Lamprecht Kind Kirchhoff Hugo Moltke
Nestroy Marie de France
Nietzsche Nansen Laotse Ipsen Liebknecht
Marx Lassalle Gorki Klett Ringelnatz
von Ossietzky May Leibniz Irving
vom Stein Lawrence
Petalozzi Platon Knigge
Sachs Pückler Michelangelo Kock Kafka
Poe Liebermann Korolenko
de Sade Praetorius Mistral Zetkin

The publishing house tredition has created the series **TREDITION CLASSICS**. It contains classical literature works from over two thousand years. Most of these titles have been out of print and off the bookstore shelves for decades.

The book series is intended to preserve the cultural legacy and to promote the timeless works of classical literature. As a reader of a **TREDITION CLASSICS** book, the reader supports the mission to save many of the amazing works of world literature from oblivion.

The symbol of **TREDITION CLASSICS** is Johannes Gutenberg (1400 – 1468), the inventor of movable type printing.

With the series, tredition intends to make thousands of international literature classics available in printed format again – worldwide.

All books are available at book retailers worldwide in paperback and in hardcover. For more information please visit: www.tredition.com

tredition was established in 2006 by Sandra Latusseck and Soenke Schulz. Based in Hamburg, Germany, tredition offers publishing solutions to authors and publishing houses, combined with worldwide distribution of printed and digital book content. tredition is uniquely positioned to enable authors and publishing houses to create books on their own terms and without conventional manufacturing risks.

For more information please visit: www.tredition.com

An Introductory Course of Quantitative Chemical Analysis With Explanatory Notes

Henry P. Talbot

Imprint

This book is part of the TREDITION CLASSICS series.

Author: Henry P. Talbot
Cover design: toepferschumann, Berlin (Germany)

Publisher: tredition GmbH, Hamburg (Germany)
ISBN: 978-3-8495-1199-9

www.tredition.com
www.tredition.de

Copyright:
The content of this book is sourced from the public domain.

The intention of the TREDITION CLASSICS series is to make world literature in the public domain available in printed format. Literary enthusiasts and organizations worldwide have scanned and digitally edited the original texts. tredition has subsequently formatted and redesigned the content into a modern reading layout. Therefore, we cannot guarantee the exact reproduction of the original format of a particular historic edition. Please also note that no modifications have been made to the spelling, therefore it may differ from the orthography used today.

PREFACE

This Introductory Course of Quantitative Analysis has been prepared to meet the needs of students who are just entering upon the subject, after a course of qualitative analysis. It is primarily intended to enable the student to work successfully and intelligently without the necessity for a larger measure of personal assistance and supervision than can reasonably be given to each member of a large class. To this end the directions are given in such detail that there is very little opportunity for the student to go astray; but the manual is not, the author believes, on this account less adapted for use with small classes, where the instructor, by greater personal influence, can stimulate independent thought on the part of the pupil.

The method of presentation of the subject is that suggested by Professor A.A. Noyes' excellent manual of Qualitative Analysis. For each analysis the procedure is given in considerable detail, and this is accompanied by explanatory notes, which are believed to be sufficiently expanded to enable the student to understand fully the underlying reason for each step prescribed. The use of the book should, nevertheless, be supplemented by classroom instruction, mainly of the character of recitations, and the student should be taught to consult larger works. The general directions are intended to emphasize those matters upon which the beginner in quantitative analysis must bestow special care, and to offer helpful suggestions. The student can hardly be expected to appreciate the force of all the statements contained in these directions, or, indeed, to retain them all in the memory after a single reading; but the instructor, by frequent reference to special paragraphs, as suitable occasion presents itself, can soon render them familiar to the student.

The analyses selected for practice are those comprised in the first course of quantitative analysis at the Massachusetts Institute of Technology, and have been chosen, after an experience of years, as affording the best preparation for more advanced work, and as satisfactory types of gravimetric and volumetric methods. From the latter point of view, they also seem to furnish the best insight into quantitative analysis for those students who can devote but a limited time to the subject, and who may never extend their study beyond the field covered by this manual. The author has had opportunity to test the efficiency of the course for use with such students, and has found the results satisfactory.

In place of the usual custom of selecting simple salts as material for preliminary practice, it has been found advantageous to substitute, in most instances, approximately pure samples of appropriate minerals or industrial products. The difficulties are not greatly enhanced, while the student gains in practical experience.

The analytical procedures described in the following pages have been selected chiefly with reference to their usefulness in teaching the subject, and with the purpose of affording as wide a variety of processes as is practicable within an introductory course of this character. The scope of the manual precludes any extended attempt to indicate alternative procedures, except through general references to larger works on analytical chemistry. The author is indebted to the standard works for many suggestions for which it is impracticable to make specific acknowledgment; no considerable credit is claimed by him for originality of procedure.

For many years, as a matter of convenience, the classes for which this text was originally prepared were divided, one part beginning with gravimetric processes and the other with volumetric analyses. After a careful review of the experience thus gained the conclusion has been reached that volumetric analysis offers the better approach to the subject. Accordingly the arrangement of the present (the sixth) edition of this manual has been changed to introduce volumetric procedures first. Teachers who are familiar with earlier editions will, however, find that the order of presentation of the material under the various divisions is nearly the same as that previously followed, and those who may still prefer to begin the course of instruction with gravimetric processes will, it is believed, be able to follow that order without difficulty.

Procedures for the determination of sulphur in insoluble sulphates, for the determination of copper in copper ores by iodometric methods, for the determination of iron by permanganate in hydrochloric acid solutions, and for the standardization of potassium permanganate solutions using sodium oxalate as a standard, and of thiosulphate solutions using copper as a standard, have been added. The determination of silica in silicates decomposable by acids, as a separate procedure, has been omitted.

The explanatory notes have been rearranged to bring them into closer association with the procedures to which they relate. The number of problems has been considerably increased.

The author wishes to renew his expressions of appreciation of the kindly reception accorded the earlier editions of this manual. He has received helpful suggestions from so many of his colleagues within the Institute, and friends elsewhere, that his sense of obligation must be expressed to them collectively. He is under special obligations to Professor L.F. Hamilton for assistance in the preparation of the present edition.

HENRY P. TALBOT

!Massachusetts Institute of Technology, September, 1921!.

CONTENTS

PART I. INTRODUCTION

SUBDIVISIONS OF ANALYTICAL CHEMISTRY
GENERAL DIRECTIONS
  Accuracy and Economy of Time; Notebooks; Reagents; Wash-bottles;
  Transfer of Liquids

PART II. VOLUMETRIC ANALYSIS

GENERAL DISCUSSION
  Subdivisions; The Analytical Balance; Weights; Burettes;
  Calibration of Measuring Devices
GENERAL DIRECTIONS
  Standard and Normal Solutions

  !I. Neutralization Methods!

ALKALIMETRY AND ACIDIMETRY
  Preparation and Standardization of Solutions; Indicators
STANDARDIZATION OF HYDROCHLORIC ACID
DETERMINATION OF TOTAL ALKALINE STRENGTH OF SODA ASH
DETERMINATION OF ACID STRENGTH OF OXALIC ACID

  !II. Oxidation Processes!

GENERAL DISCUSSION BICHROMATE PROCESS FOR THE DETERMINATION OF IRON DETERMINATION OF IRON IN LIMONITE BY THE BICHROMATE PROCESS DETERMINATION OF CHROMIUM IN CHROME IRON ORE PERMANGA-

NATE PROCESS FOR THE DETERMINATION OF IRON DETERMINATION OF IRON IN LIMONITE BY THE PERMANGANATE PROCESS DETERMINATION OF IRON IN LIMONITE BY THE ZIMMERMANN-REINHARDT PROCESS DETERMINATION OF THE OXIDIZING POWER OF PYROLUSITE IODIMETRY DETERMINATION OF COPPER IN ORES DETERMINATION OF ANTIMONY IN STIBNITE CHLORIMETRY DETERMINATION OF AVAILABLE CHLORINE IN BLEACHING POWDER

!III. Precipitation Methods!

DETERMINATION OF SILVER BY THE THIOCYANATE PROCESS

PART III. GRAVIMETRIC ANALYSIS

GENERAL DIRECTIONS
  Precipitation; Funnels and Filters; Filtration and Washing of
  Precipitates; Desiccators; Crucibles and their Preparation
  for Use; Ignition of Precipitates
DETERMINATION OF CHLORINE IN SODIUM CHLORIDE
DETERMINATION OF IRON AND OF SULPHUR IN FERROUS AMMONIUM SULPHATE
DETERMINATION OF SULPHUR IN BARIUM SULPHATE
DETERMINATION OF PHOSPHORIC ANHYDRIDE IN APATITE
ANALYSIS OF LIMESTONE
  Determination of Moisture; Insoluble Matter and Silica; Ferric
  Oxide and Alumina; Calcium; Magnesium; Carbon Dioxide
ANALYSIS OF BRASS
  Electrolytic Separations; Determination of Lead, Copper, Iron
  and Zinc.
DETERMINATION OF SILICA IN SILICATES

PART IV. STOICHIOMETRY

SOLUTIONS OF TYPICAL PROBLEMS PROBLEMS

APPENDIX

ELECTROLYTIC DISSOCIATION THEORY FOLDING OF A FILTER PAPER SAMPLE NOTEBOOK PAGES STRENGTH OF REAGENTS DENSITIES AND VOLUMES OF WATER CORRECTIONS FOR CHANGE OF TEMPERATURE OF STANDARD SOLUTIONS ATOMIC WEIGHTS LOGARITHM TABLES

QUANTITATIVE CHEMICAL ANALYSIS

PART I

INTRODUCTION

SUBDIVISIONS OF ANALYTICAL CHEMISTRY

A complete chemical analysis of a body of unknown composition involves the recognition of its component parts by the methods of !qualitative analysis!, and the determination of the proportions in which these components are present by the processes of !quantitative analysis!. A preliminary qualitative examination is generally indispensable, if intelligent and proper provisions are to be made for the separation of the various constituents under such conditions as will insure accurate quantitative estimations.

It is assumed that the operations of qualitative analysis are familiar to the student, who will find that the reactions made use of in quantitative processes are frequently the same as those employed in qualitative analyses with respect to both precipitation and systematic separation from interfering substances; but it should be noted that the conditions must now be regulated with greater care, and in such a manner as to insure the most complete separation possible. For example, in the qualitative detection of sulphates by precipitation as barium sulphate from acid solution it is not necessary, in most instances, to take into account the solubility of the sulphate in hydrochloric acid, while in the quantitative determination of sulphates by this reaction this solubility becomes an important consideration. The operations of qualitative analysis are, therefore, the more accurate the nearer they are made to conform to quantitative conditions.

The methods of quantitative analysis are subdivided, according to their nature, into those of !gravimetric analysis, volumetric analysis!, and !colorimetric analysis!. In !gravimetric! processes the constituent to be determined is sometimes isolated in elementary form, but more commonly in the form of some compound possessing a well-established and definite composition, which can be readily and completely separated, and weighed either directly or after ignition. From the weight of this substance and its known composition, the amount of the constituent in question is determined.

In !volumetric! analysis, instead of the final weighing of a definite body, a well-defined reaction is caused to take place, wherein the reagent is added from an apparatus so designed that the volume of the solution employed to complete the reaction can be accurately measured. The strength of this solution (and hence its value for the reaction in question) is accurately known, and the volume employed serves, therefore, as a measure of the substance acted upon. An example will make clear the distinction between these two types of analysis. The percentage of chlorine in a sample of sodium chloride may be determined by dissolving a weighed amount of the chloride in water and precipitating the chloride ions as silver chloride, which is then separated by filtration, ignited, and weighed (a !gravimetric! process); or the sodium chloride may be dissolved in water, and a solution of silver nitrate containing an accurately known amount of the silver salt in each cubic centimeter may be cautiously added from a measuring device called a burette until precipitation is complete, when the amount of chlorine may be calculated from the number of cubic centimeters of the silver nitrate solution involved in the reaction. This is a !volumetric! process, and is equivalent to weighing without the use of a balance.

Volumetric methods are generally more rapid, require less apparatus, and are frequently capable of greater accuracy than gravimetric methods. They are particularly useful when many determinations of the same sort are required.

In !colorimetric! analyses the substance to be determined is converted into some compound which imparts to its solutions a distinct color, the intensity of which must vary in direct proportion to the amount of the compound in the solution. Such solutions are compared with respect to depth of color with standard solutions containing known amounts of the colored compound, or of other similar color-producing substance which has been found acceptable as a color standard. Colorimetric methods are, in general, restricted to the determinations of very small quantities, since only in dilute solutions are accurate comparisons of color possible.

GENERAL DIRECTIONS

The following paragraphs should be read carefully and thoughtfully. A prime essential for success as an analyst is attention to details and the avoidance of all conditions which could destroy, or even lessen, confidence in the analyses when completed. The suggestions here given are the outcome of much experience, and their adoption will tend to insure permanently work of a high grade, while neglect of them will often lead to disappointment and loss of time.

ACCURACY AND ECONOMY OF TIME

The fundamental conception of quantitative analysis implies a necessity for all possible care in guarding against loss of material or the introduction of foreign matter. The laboratory desk, and all apparatus, should be scrupulously neat and clean at all times. A sponge should always be ready at hand, and desk and filter-stands should be kept dry and in good order. Funnels should never be allowed to drip upon the base of the stand. Glassware should always be wiped with a clean, lintless towel just before use. All filters and solutions should be covered to protect them from dust, just as far as is practicable, and every drop of solution or particle of precipitate must be regarded as invaluable for the success of the analysis.

An economical use of laboratory hours is best secured by acquiring a thorough knowledge of the character of the work to be done before undertaking it, and then by so arranging the work that no time shall be wasted during the evaporation of liquids and like time-consuming operations. To this end the student should read thoughtfully not only the !entire! procedure, but the explanatory notes as well, before any step is taken in the analysis. The explanatory notes furnish, in general, the reasons for particular steps or precautions, but they also occasionally contain details of manipulation not incorporated, for various reasons, in the procedure. These notes follow the procedures at frequent intervals, and the exact points to which they apply are indicated by references. The student should realize that a !failure to study the notes will inevitably lead to mistakes, loss of time, and an inadequate understanding of the subject!.

All analyses should be made in duplicate, and in general a close agreement of results should be expected. It should, however, be remembered that a close concordance of results in "check analyses" is not conclusive evidence of the accuracy of those results, although the probability of their accuracy is, of course, considerably enhanced. The satisfaction in obtaining "check results" in such analyses must never be allowed to interfere with the critical examination of the procedure employed, nor must they ever be regarded as in any measure a substitute for absolute truth and accuracy.

In this connection it must also be emphasized that only the operator himself can know the whole history of an analysis, and only he can know whether his work is worthy of full confidence. No work should be continued for a moment after such confidence is lost, but should be resolutely discarded as soon as a cause for distrust is fully established. The student should, however, determine to put forth his best efforts in each analysis; it is well not to be too ready to condone failures and to "begin again," as much time is lost in these fruitless attempts. Nothing less than !absolute integrity! is or can be demanded of a quantitative analyst, and any disregard of this principle, however slight, is as fatal to success as lack of chemical knowledge or inaptitude in manipulation can possibly be.

NOTEBOOKS

Notebooks should contain, beside the record of observations, descriptive notes. All records of weights should be placed upon the right-hand page, while that on the left is reserved for the notes, calculations of factors, or the amount of reagents required.

The neat and systematic arrangement of the records of analyses is of the first importance, and is an evidence of careful work and an excellent credential. Of two notebooks in which the results may be, in fact, of equal value as legal evidence, that one which is neatly arranged will carry with it greater weight.

All records should be dated, and all observations should be recorded at once in the notebook. The making of records upon loose paper is a practice to be deprecated, as is also that of copying original entries into a second notebook. The student should accustom himself to orderly entries at the time of observation. Several sample pages of systematic records are to be found in the Appendix. These are based upon experience; but other arrangements, if clear and orderly, may prove equally serviceable. The student is advised to follow the sample pages until he is in a position to plan out a system of his own.

REAGENTS

The habit of carefully testing reagents, including distilled water, cannot be too early acquired or too constantly practiced; for, in spite of all reasonable precautionary measures, inferior chemicals will occasionally find their way into the stock room, or errors will be made in filling reagent bottles. The student should remember that while there may be others who share the responsibility for the purity of materials in the laboratory of an institution, the responsibility will later be one which he must individually assume.

The stoppers of reagent bottles should never be laid upon the desk, unless upon a clean watch-glass or paper. The neck and mouth of all such bottles should be kept scrupulously clean, and care taken that no confusion of stoppers occurs.

WASH-BOTTLES

Wash-bottles for distilled water should be made from flasks of about 750 cc. capacity and be provided with gracefully bent tubes, which should not be too long. The jet should be connected with the tube entering the wash-bottle by a short piece of rubber tubing in such a way as to be flexible, and should deliver a stream about one millimeter in diameter. The neck of the flask may be wound with cord, or covered with wash-leather, for greater comfort when hot water is used. It is well to provide several small wash-bottles for liquids other than distilled water, which should invariably be clearly labeled.

TRANSFER OF LIQUIDS

Liquids should never be transferred from one vessel to another, nor to a filter, without the aid of a stirring rod held firmly against the side or lip of the vessel. When the vessel is provided with a lip it is not usually necessary to use other means to prevent the loss of liquid by running down the side; whenever loss seems imminent a !very thin! layer of vaseline, applied with the finger to the edge of the vessel, will prevent it. The stirring rod down which the liquid runs should never be drawn upward in such a way as to allow the solution to collect on the under side of the rim or lip of a vessel.

The number of transfers of liquids from one vessel to another during an analysis should be as small as possible to avoid the risk of slight losses. Each vessel must, of course, be completely washed to insure the transfer of all material; but it should be remembered that this can be accomplished better by the use of successive small portions of wash-water (perhaps 5-10 cc.), if each wash-water is allowed to drain away for a few seconds, than by the addition of large amounts which unnecessarily increase the volume of the solutions, causing loss of time in subsequent filtrations or evaporations.

All stirring rods employed in quantitative analyses should be rounded at the ends by holding them in the flame of a burner until they begin to soften. If this is not done, the rods will scratch the inner surface of beakers, causing them to crack on subsequent heating.

EVAPORATION OF LIQUIDS

The greatest care must be taken to prevent loss of solutions during processes of evaporation, either from too violent ebullition, from evaporation to dryness and spattering, or from the evolution of gas during the heating. In general, evaporation upon the steam bath is to be preferred to other methods on account of the impossibility of loss by spattering. If the steam baths are well protected from dust, solutions should be left without covers during evaporation; but solutions which are boiled upon the hot plate, or from which gases are escaping, should invariably be covered. In any case a watch-glass may be supported above the vessel by means of a glass triangle, or other similar device, and the danger of loss of material or contamination by dust thus be avoided. It is obvious that evaporation is promoted by the use of vessels which admit of the exposure of a broad surface to the air.

Liquids which contain suspended matter (precipitates) should always be cautiously heated, since the presence of the solid matter is frequently the occasion of violent "bumping," with consequent risk to apparatus and analysis.

PART II

VOLUMETRIC ANALYSIS

The processes of volumetric analysis are, in general, simpler than those of gravimetric analysis and accordingly serve best as an introduction to the practice of quantitative analysis. For their execution there are required, first, an accurate balance with which to weigh the material for analysis; second, graduated instruments in which to measure the volume of the solutions employed; third, standard solutions, that is, solutions the value of which is accurately known; and fourth, indicators, which will furnish accurate evidence of the point at which the desired reaction is completed. The nature of the indicators employed will be explained in connection with the different analyses.

The process whereby a !standard solution! is brought into reaction is called !titration!, and the point at which the reaction is exactly completed is called the !end-point!. The !indicator! should show the !end-point! of the !titration!. The volume of the standard solution used then furnishes the measure of the substance to be determined as truly as if that substance had been separated and weighed.

The processes of volumetric analysis are easily classified, according to their character, into:

I. NEUTRALIZATION METHODS; such, for example, as those of acidimetry and alkalimetry.

II. OXIDATION PROCESSES; as exemplified in the determination of ferrous iron by its oxidation with potassium bichromate.

III. PRECIPITATION METHODS; of which the titration for silver with potassium thiocyanate solution is an illustration.

From a somewhat different standpoint the methods in each case may be subdivided into (a) DIRECT METHODS, in which the substance to be measured is directly determined by titration to an end-point with a standard solution; and (b) INDIRECT METHODS, in which the substance itself is not measured, but a quantity of reagent is added which is known to be an excess with respect to a specific reaction, and the unused excess determined by titration. Examples of the latter class will be pointed out as they occur in the procedures.

MEASURING INSTRUMENTS

THE ANALYTICAL BALANCE

For a complete discussion of the physical principles underlying the construction and use of balances, and the various methods of weighing, the student is referred to larger manuals of Quantitative Analysis, such as those of Fresenius, or Treadwell-Hall, and particularly to the admirable discussion of this topic in Morse's !Exercises in Quantitative Chemistry!.

The statements and rules of procedure which follow are sufficient for the intelligent use of an analytical balance in connection with processes prescribed in this introductory manual. It is, however, imperative that the student should make himself familiar with these essential features of the balance, and its use. He should fully realize that the analytical balance is a delicate instrument which will render excellent service under careful treatment, but such treatment is an essential condition if its accuracy is to be depended upon. He should also understand that no set of rules, however complete, can do away with the necessity for a sense of personal responsibility, since by carelessness he can render inaccurate not only his own analyses, but those of all other students using the same balance.

Before making any weighings the student should seat himself before a balance and observe the following details of construction:

1. The balance case is mounted on three brass legs, which should preferably rest in glass cups, backed with rubber to prevent slipping. The front legs are adjustable as to height and are used to level the balance case; the rear leg is of permanent length.

2. The front of the case may be raised to give access to the balance. In some makes doors are provided also at the ends of the balance case.

3. The balance beam is mounted upon an upright in the center of the case on the top of which is an inlaid agate plate. To the center of the beam there is attached a steel or agate knife-edge on which the beam oscillates when it rests on the agate plate.

4. The balance beam, extending to the right and left, is graduated along its upper edge, usually on both sides, and has at its extremities two agate or steel knife-edges from which are suspended stirrups. Each of these stirrups has an agate plate which, when the balance is in action, rests upon the corresponding knife-edge of the beam. The balance pans are suspended from the stirrups.

5. A pointer is attached to the center of the beam, and as the beam oscillates this pointer moves in front of a scale near the base of the post.

6. At the base of the post, usually in the rear, is a spirit-level.

7. Within the upright is a mechanism, controlled by a knob at the front of the balance case, which is so arranged as to raise the entire beam slightly above the level at which the knife-edges are in contact with the agate plates. When the balance is not in use the beam must be supported by this device since, otherwise, the constant jarring to which a balance is inevitably subjected, will soon dull the knife-edges, and lessen the sensitiveness of the balance.

8. A small weight, or bob, is attached to the pointer (or sometimes to the beam) by which the center of gravity of the beam and its attachments may be regulated. The center of gravity must lie very slightly below the level of the agate plates to secure the desired sensitiveness of the balance. This is provided for when the balance is set up and very rarely requires alteration. The student should never attempt to change this adjustment.

9. Below the balance pans are two pan-arrests operated by a button from the front of the case. These arrests exert a very slight upward pressure upon the pans and minimize the displacement of the beam when objects or weights are being placed upon the pans.

10. A movable rod, operated from one end of the balance case, extends over the balance beam and carries a small wire weight, called a rider. By means of this rod the rider can be placed upon any desired division of the scale on the balance beam. Each numbered division on the beam corresponds to one milligram, and the use of the rider obviates the placing of very small fractional weights on the balance pan.

If a new rider is purchased, or an old one replaced, care must be taken that its weight corresponds to the graduations on the beam of the balance on which it is to be used. The weight of the rider in milligrams must be equal to the number of large divisions (5, 6, 10, or 12) between the central knife-edge and the knife-edge at the end of the beam. It should be noted that on some balances the last division bears no number. Each new rider should be tested against a 5 or 10-milligram weight.

In some of the most recent forms of the balance a chain device replaces the smaller weights and the use of the rider as just described.

Before using a balance, it is always best to test its adjustment. This is absolutely necessary if the balance is used by several workers; it is always a wise precaution under any conditions. For this purpose, brush off the balance pans with a soft camel's hair brush. Then note (1) whether the balance is level; (2) that the mechanism for raising and lowering the beams works smoothly; (3) that the pan-arrests touch the pans when the beam is lowered; and (4) that the needle swings equal distances on either side of the zero-point when set in motion without any load on the pans. If the latter condition is not fulfilled, the balance should be adjusted by means of the adjusting screw at the end of the beam unless the variation is not more than one division on the scale; it is often better to make a proper allowance for this small zero error than to disturb the balance by an attempt at correction. Unless a student thoroughly understands the construction of a balance he should never attempt to make adjustments, but should apply to the instructor in charge.

The object to be weighed should be placed on the left-hand balance pan and the weights upon the right-hand pan. Every substance which could attack the metal of the balance pan should be weighed upon a watch-glass, and all objects must be dry and cold. A warm body gives rise to air currents which vitiate the accuracy of the weighing.

The weights should be applied in the order in which they occur in the weight-box (not at haphazard), beginning with the largest weight which is apparently required. After a weight has been placed upon the pan the beam should be lowered upon its knife-edges, and, if necessary, the pan-arrests depressed. The movement of the pointer will then indicate whether the weight applied is too great or too small. When the weight has been ascertained, by the successive addition of small weights, to the nearest 5 or 10 milligrams, the weighing is completed by the use of the rider. The correct weight is that which causes the pointer to swing an equal number of divisions to the right and left of the zero-point, when the pointer traverses not less than five divisions on either side.

The balance case should always be closed during the final weighing, while the rider is being used, to protect the pans from the effect of air currents.

Before the final determination of an exact weight the beam should always be lifted from the knife-edges and again lowered into place, as it frequently happens that the scale pans are, in spite of the pan-arrests, slightly twisted by the impact of the weights, the beam being thereby virtually lengthened or shortened. Lifting the beam restores the proper alignment.

The beam should never be set in motion by lowering it forcibly upon the knife-edges, nor by touching the pans, but rather by lifting the rider (unless the balance be provided with some of the newer devices for the purpose), and the swing should be arrested only when the needle approaches zero on the scale, otherwise the knife-edges become dull. For the same reason the beam should never be left upon its knife-edges, nor should weights be removed from or placed on the pans without supporting the beam, except in the case of the small fractional weights.

When the process of weighing has been completed, the weight should be recorded in the notebook by first noting the vacant spaces in the weight-box, and then checking the weight by again noting the weights as they are removed from the pan. This practice will often detect and avoid errors. It is obvious that the weights should always be returned to their proper places in the box, and be handled only with pincers.

It should be borne in mind that if the mechanism of a balance is deranged or if any substance is spilled upon the pans or in the balance case, the damage should be reported at once. In many instances serious harm can be averted by prompt action when delay might ruin the balance.

Samples for analysis are commonly weighed in small tubes with cork stoppers. Since the stoppers are likely to change in weight from the varying amounts of moisture absorbed from the atmosphere, it is necessary to confirm the recorded weight of a tube which has been unused for some time before weighing out a new portion of substance from it.

WEIGHTS

The sets of weights commonly used in analytical chemistry range from 20 grams to 5 milligrams. The weights from 20 grams to 1 gram are usually of brass, lacquered or gold plated. The fractional weights are of German silver, gold, platinum or aluminium. The rider is of platinum or aluminium wire.

The sets of weights purchased from reputable dealers are usually sufficiently accurate for analytical work. It is not necessary that such a set should be strictly exact in comparison with the absolute standard of weight, provided they are relatively correct among themselves, and provided the same set of weights is used in all weighings made during a given analysis. The analyst should assure himself that the weights in a set previously unfamiliar to him are relatively correct by a few simple tests. For example, he should make sure that in his set two weights of the same denomination (i.e., two 10-gram weights, or the two 100-milligram weights) are actually equal and interchangeable, or that the 500-milligram weight is equal to the sum of the 200, 100, 100, 50, 20, 20 and 10-milligram weights combined, and so on. If discrepancies of more than a few tenths of a milligram (depending upon the total weight involved) are found, the weights should be returned for correction. The rider should also be compared with a 5 or 10-milligram weight.

In an instructional laboratory appreciable errors should be reported to the instructor in charge for his consideration.

When the highest accuracy is desired, the weights may be calibrated and corrections applied. A calibration procedure is described in a paper by T.W. Richards, !J. Am. Chem. Soc.!, 22, 144, and in many large text-books.

Weights are inevitably subject to corrosion if not properly protected at all times, and are liable to damage unless handled with great care. It is obvious that anything which alters the weight of a single piece in an analytical set will introduce an error in every weighing made in which that piece is used. This source of error is often extremely obscure and difficult to detect. The only safeguard against such errors is to be found in scrupulous care in handling and protection on the part of the analyst, and an equal insistence that if several analysts use the same set of weights, each shall realize his responsibility for the work of others as well as his own.

BURETTES

A burette is made from a glass tube which is as uniformly cylindrical as possible, and of such a bore that the divisions which are etched upon its surface shall correspond closely to actual contents.

The tube is contracted at one extremity, and terminates in either a glass stopcock and delivery-tube, or in such a manner that a piece of rubber tubing may be firmly attached, connecting a delivery-tube of glass. The rubber tubing is closed by means of a glass bead. Burettes of the latter type will be referred to as "plain burettes."

The graduations are usually numbered in cubic centimeters, and the latter are subdivided into tenths.

One burette of each type is desirable for the analytical procedures which follow.

PREPARATION OF A BURETTE FOR USE

The inner surface of a burette must be thoroughly cleaned in order that the liquid as drawn out may drain away completely, without leaving drops upon the sides. This is best accomplished by treating the inside of the burette with a warm solution of chromic acid in concentrated sulphuric acid, applied as follows: If the burette is of the "plain" type, first remove the rubber tip and force the lower end of the burette into a medium-sized cork stopper. Nearly fill the burette with the chromic acid solution, close the upper end with a cork stopper and tip the burette backward and forward in such a way as to bring the solution into contact with the entire inner surface. Remove the stopper and pour the solution into a stock bottle to be kept for further use, and rinse out the burette with water several times. Unless the water then runs freely from the burette without leaving drops adhering to the sides, the process must be repeated (Note 1).

If the burette has a glass stopcock, this should be removed after the cleaning and wiped, and also the inside of the ground joint. The surface of the stopcock should then be smeared with a thin coating of vaseline and replaced. It should be attached to the burette by means of a wire, or elastic band, to lessen the danger of breakage.

Fill the burettes with distilled water, and allow the water to run out through the stopcock or rubber tip until convinced that no air bubbles are inclosed (Note 2). Fill the burette to a point above the zero-point and draw off the water until the meniscus is just below that mark. It is then ready for calibration.

[Note 1: The inner surface of the burette must be absolutely clean if the liquid is to run off freely. Chromic acid in sulphuric acid is usually found to be the best cleansing agent, but the mixture must be warm and concentrated. The solution can be prepared by pouring over a few crystals of potassium bichromate a little water and then adding concentrated sulphuric acid.]

[Note 2: It is always necessary to insure the absence of air bubbles in the tips or stopcocks. The treatment described above will usually accomplish this, but, in the case of plain burettes it is sometimes better to allow a little of the liquid to flow out of the tip while it is bent upwards. Any air which may be entrapped then rises with the liquid and escapes.

If air bubbles escape during subsequent calibration or titration, an error is introduced which vitiates the results.]

READING OF A BURETTE

All liquids when placed in a burette form what is called a meniscus at their upper surfaces. In the case of liquids such as water or aqueous solutions this meniscus is concave, and when the liquids are transparent accurate readings are best obtained by observing the position on the graduated scales of the lowest point of the meniscus. This can best be done as follows: Wrap around the burette a piece of colored paper, the straight, smooth edges of which are held evenly together with the colored side next to the burette (Note 1). Hold the paper about two small divisions below the meniscus and raise or lower the level of the eyes until the edge of the paper at the back of the burette is just hidden from the eye by that in front (Note 2). Note the position of the lowest point of the curve of the meniscus, estimating the tenths of the small divisions, thus reading its position to hundredths of a cubic centimeter.

[Note 1: The ends of the colored paper used as an aid to accurate readings may be fastened together by means of a gummed label. The paper may then remain on the burette and be ready for immediate use by sliding it up or down, as required.]

[Note 2: To obtain an accurate reading the eye must be very nearly on a level with the meniscus. This is secured by the use of the paper as described. The student should observe by trial how a reading is affected when the meniscus is viewed from above or below.

The eye soon becomes accustomed to estimating the tenths of the divisions. If the paper is held as directed, two divisions below the meniscus, one whole division is visible to correct the judgment. It is not well to attempt to bring the meniscus exactly to a division mark on the burette. Such readings are usually less accurate than those in which the tenths of a division are estimated.]

CALIBRATION OF GLASS MEASURING DEVICES

If accuracy of results is to be attained, the correctness of all measuring instruments must be tested. None of the apparatus offered for sale can be implicitly relied upon except those more expensive instruments which are accompanied by a certificate from the !National Bureau of Standards! at Washington, or other equally authentic source.

The bore of burettes is subject to accidental variations, and since the graduations are applied by machine without regard to such variations of bore, local errors result.

The process of testing these instruments is called !calibration!. It is usually accomplished by comparing the actual weight of water contained in the instrument with its apparent volume.

There is, unfortunately, no uniform standard of volume which has been adopted for general use in all laboratories. It has been variously proposed to consider the volume of 1000 grams of water at 4°, 15.5°, 16°, 17.5°, and even 20°C., as a liter for practical purposes, and to consider the cubic centimeter to be one one-thousandth of that volume. The true liter is the volume of 1000 grams of water at 4°C.; but this is obviously a lower temperature than that commonly found in laboratories, and involves the constant use of corrections if taken as a laboratory standard. Many laboratories use 15.5°C. (60° F.) as the working standard. It is plain that any temperature which is deemed most convenient might be chosen for a particular laboratory, but it cannot be too strongly emphasized that all measuring instruments, including burettes, pipettes, and flasks, should be calibrated at that temperature in order that the contents of each burette, pipette, etc., shall be comparable with that of every other instrument, thus permitting general interchange and substitution. For example, it is obvious that if it is desired to remove exactly 50 cc. from a solution which has been diluted to 500 cc. in a graduated flask, the 50 cc. flask or pipette used to remove the fractional portion must give a correct reading at the same temperature as the 500 cc. flask. Similarly, a burette used for the titration of the 50 cc. of solution removed should be calibrated under the same conditions as the measuring flasks or pipettes employed with it.

The student should also keep constantly in mind the fact that all volumetric operations, to be exact, should be carried out as nearly at a constant temperature as is practicable. The spot selected for such work should therefore be subject to a minimum of temperature variations, and should have as nearly the average temperature of the laboratory as is possible. In all work, whether of calibration, standardization, or analysis, the temperature of the liquids employed must be taken into account, and if the temperature of these liquids varies more than 3° or 4° from the standard temperature chosen for the laboratory, corrections must be applied for errors due to expansion or contraction, since volumes of a liquid measured at different times are comparable only under like conditions as to temperature. Data to be used for this purpose are given in the Appendix. Neglect of this correction is frequently an avoidable source of error and annoyance in otherwise excellent work. The temperature of all solutions at the time of standardization should be recorded to facilitate the application of temperature corrections, if such are necessary at any later time.

CALIBRATION OF THE BURETTES

Two burettes, one at least of which should have a glass stopper, are required throughout the volumetric work. Both burettes should be calibrated by the student to whom they are assigned.

PROCEDURE.—Weigh a 50 cc., flat-bottomed flask (preferably a light-weight flask), which must be dry on the outside, to the nearest centigram. Record the weight in the notebook. (See Appendix for suggestions as to records.) Place the flask under the burette and draw out into it about 10 cc. of water, removing any drop on the tip by touching it against the inside of the neck of the flask. Do not attempt to stop exactly at the 10 cc. mark, but do not vary more than 0.1 cc. from it. Note the time, and at the expiration of three minutes (or longer) read the burette accurately, and record the reading in the notebook (Note 1). Meanwhile weigh the flask and water to centigrams and record its weight (Note 2). Draw off the liquid from 10 cc. to about 20 cc. into the same flask without emptying it; weigh, and at the expiration of three minutes take the reading, and so on throughout the length of the burette. When it is completed, refill the burette and check the first calibration.

The differences in readings represent the apparent volumes, the differences in weights the true volumes. For example, if an apparent volume of 10.05 cc. is found to weigh 10.03 grams, it may be assumed with sufficient accuracy that the error in that 10 cc. amounts to -0.02 cc., or -0.002 for each cubic centimeter (Note 3).

In the calculation of corrections the temperature of the water must be taken into account, if this varies more than 4°C. from the laboratory standard temperature, consulting the table of densities of water in the Appendix.

From the final data, plot the corrections to be applied so that they may be easily read for each cubic centimeter throughout the burette. The total correction at each 10 cc. may also be written on the burette with a diamond, or etching ink, for permanence of record.

[Note 1: A small quantity of liquid at first adheres to the side of even a clean burette. This slowly unites with the main body of liquid, but requires an appreciable time. Three minutes is a sufficient interval, but not too long, and should be adopted in every instance throughout the whole volumetric practice before final readings are recorded.]

[Note 2: A comparatively rough balance, capable of weighing to centigrams, is sufficiently accurate for use in calibrations, for a moment's reflection will show that it would be useless to weigh the water with an accuracy greater than that of the readings taken on the burette. The latter cannot exceed 0.01 cc. in accuracy, which corresponds to 0.01 gram.

The student should clearly understand that !all other weighings!, except those for calibration, should be made accurately to 0.0001 gram, unless special directions are given to the contrary.

Corrections for temperature variations of less than 4°C. are negligible, as they amount to less than 0.01 gram for each 10 grams of water withdrawn.]

[Note 3: Should the error discovered in any interval of 10 cc. on the burette exceed 0.10 cc., it is advisable to weigh small portions (even 1 cc.) to locate the position of the variation of bore in the tube rather than to distribute the correction uniformly over the corresponding 10 cc. The latter is the usual course for small corrections, and it is convenient to calculate the correction corresponding to each cubic centimeter and to record it in the form of a table or calibration card, or to plot a curve representing the values.

Burettes may also be calibrated by drawing off the liquid in successive portions through a 5 cc. pipette which has been accurately calibrated, as a substitute for weighing. If many burettes are to be tested, this is a more rapid method.]

PIPETTES

A !pipette! may consist of a narrow tube, in the middle of which is blown a bulb of a capacity a little less than that which it is desired to measure by the pipette; or it may be a miniature burette, without the stopcock or rubber tip at the lower extremity. In either case, the flow of liquid is regulated by the pressure of the finger on the top, which governs the admission of the air.

Pipettes are usually already graduated when purchased, but they require calibration for accurate work.

CALIBRATION OF PIPETTES

PROCEDURE.—Clean the pipette. Draw distilled water into it by sucking at the upper end until the water is well above the graduation mark. Quickly place the forefinger over the top of the tube, thus preventing the entrance of air and holding the water in the pipette. Cautiously admit a little air by releasing the pressure of the finger, and allow the level of the water to fall until the lowest point of the meniscus is level with the graduation. Hold the water at that point by pressure of the finger and then allow the water to run out from the pipette into a small tared, or weighed, beaker or flask. After a definite time interval, usually two to three minutes, touch the end of the pipette against the side of the beaker or flask to remove any liquid adhering to it (Note 1). The increase in weight of the flask in grams represents the volume of the water in cubic centimeters delivered by the pipette. Calculate the necessary correction.

[Note 1: A definite interval must be allowed for draining, and a definite practice adopted with respect to the removal of the liquid which collects at the end of the tube, if the pipette is designed to deliver a specific volume when emptied. This liquid may be removed at the end of a definite interval either by touching the side of the vessel or by gently blowing out the last drops. Either practice, when adopted, must be uniformly adhered to.]

FLASKS

!Graduated or measuring flasks! are similar to the ordinary flat-bottomed flasks, but are provided with long, narrow necks in order that slight variations in the position of the meniscus with respect to the graduation shall represent a minimum volume of liquid. The flasks must be of such a capacity that, when filled with the specified volume, the liquid rises well into the neck.

GRADUATION OF FLASKS

It is a general custom to purchase the flasks ungraduated and to graduate them for use under standard conditions selected for the laboratory in question. They may be graduated for "contents" or "delivery." When graduated for "contents" they contain a specified volume when filled to the graduation at a specified temperature, and require to be washed out in order to remove all of the solution from the flask. Flasks graduated for "delivery" will deliver the specified volume of a liquid without rinsing. A flask may, of course, be graduated for both contents and delivery by placing two graduation marks upon it.

PROCEDURE.—To calibrate a flask for !contents!, proceed as follows: Clean the flask, using a chromic acid solution, and dry it carefully outside and inside. Tare it accurately; pour water into the flask until the weight of the latter counterbalances weights on the opposite pan which equal in grams the number of cubic centimeters of water which the flask is to contain. Remove any excess of water with the aid of filter paper (Note 1). Take the flask from the balance, stopper it, place it in a bath at the desired temperature, usually 15.5° or 17.5°C., and after an hour mark on the neck with a diamond the location of the lowest point of the meniscus (Note 2). The mark may be etched upon the flask by hydrofluoric acid, or by the use of an etching ink now commonly sold on the market.

To graduate a flask which is designed to !deliver! a specified volume, proceed as follows: Clean the flask as usual and wipe all moisture from the outside. Fill it with distilled water. Pour out the water and allow the water to drain from the flask for three minutes. Counterbalance the flask with weights to the nearest centigram. Add weights corresponding in grams to the volume desired, and add distilled water to counterbalance these weights. An excess of water, or water adhering to the neck of the flask, may be removed by means of a strip of clean filter paper. Stopper the flask, place it in a bath at 15.5°C. or 17.5°C. and, after an hour, mark the location of the lowest point of the meniscus, as described above.

[Note 1: The allowable error in counterbalancing the water and weights varies with the volume of the flask. It should not exceed one ten-thousandth of the weight of water.]

[Note 2: Other methods are employed which involve the use of calibrated apparatus from which the desired volume of water may be run into the dry flask and the position of the meniscus marked directly upon it. For a description of a procedure which is most convenient when many flasks are to be calibrated, the student is referred to the !Am. Chem J.!, 16, 479.]

GENERAL DIRECTIONS FOR VOLUMETRIC ANALYSES

It cannot be too strongly emphasized that for the success of analyses uniformity of practice must prevail throughout all volumetric work with respect to those factors which can influence the accuracy of the measurement of liquids. For example, whatever conditions are imposed during the calibration of a burette, pipette, or flask (notably the time allowed for draining), must also prevail whenever the flask or burette is used.

The student should also be constantly watchful to insure parallel conditions during both standardization and analyst with respect to the final volume of liquid in which a titration takes place. The value of a standard solution is only accurate under the conditions which prevailed when it was standardized. It is plain that the standard solutions must be scrupulously protected from concentration or dilution, after their value has been established. Accordingly, great care must be taken to thoroughly rinse out all burettes, flasks, etc., with the solutions which they are to contain, in order to remove all traces of water or other liquid which could act as a diluent. It is best to wash out a burette at least three times with small portions of a solution, allowing each to run out through the tip before assuming that the burette is in a condition to be filled and used. It is, of course, possible to dry measuring instruments in a hot closet, but this is tedious and unnecessary.

To the same end, all solutions should be kept stoppered and away from direct sunlight or heat. The bottles should be shaken before use to collect any liquid which may have distilled from the solution and condensed on the sides.

The student is again reminded that variations in temperature of volumetric solutions must be carefully noted, and care should always be taken that no source of heat is sufficiently near the solutions to raise the temperature during use.

Much time may be saved by estimating the approximate volume of a standard solution which will be required for a titration (if the data are obtainable) before beginning the operation. It is then possible to run in rapidly approximately the required amount, after which it is only necessary to determine the end-point slowly and with accuracy. In such cases, however, the knowledge of the approximate amount to be required should never be allowed to influence the judgment regarding the actual end-point.

STANDARD SOLUTIONS

The strength or value of a solution for a specific reaction is determined by a procedure called !Standardization!, in which the solution is brought into reaction with a definite weight of a substance of known purity. For example, a definite weight of pure sodium carbonate may be dissolved in water, and the volume of a solution of hydrochloric acid necessary to exactly neutralize the carbonate accurately determined. From these data the strength or value of the acid is known. It is then a !standard solution!.

NORMAL SOLUTIONS

Standard solutions may be made of a purely empirical strength dictated solely by convenience of manipulation, or the concentration may be chosen with reference to a system which is applicable to all solutions, and based upon chemical equivalents. Such solutions are called !Normal Solutions! and contain such an amount of the reacting substance per liter as is equivalent in its chemical action to one gram of hydrogen, or eight grams of oxygen. Solutions containing one half, one tenth, or one one-hundredth of this quantity per liter are called, respectively, half-normal, tenth-normal, or hundredth-normal solutions.

Since normal solutions of various reagents are all referred to a common standard, they have an advantage not possessed by empirical solutions, namely, that they are exactly equivalent to each other. Thus, a liter of a normal solution of an acid will exactly neutralize a liter of a normal alkali solution, and a liter of a normal oxidizing solution will exactly react with a liter of a normal reducing solution, and so on.

Beside the advantage of uniformity, the use of normal solutions simplifies the calculations of the results of analyses. This is particularly true if, in connection with the normal solution, the weight of substance for analysis is chosen with reference to the atomic or molecular weight of the constituent to be determined. (See problem 26.)

The preparation of an !exactly! normal, half-normal, or tenth-normal solution requires considerable time and care. It is usually carried out only when a large number of analyses are to be made, or when the analyst has some other specific purpose in view. It is, however, a comparatively easy matter to prepare standard solutions which differ but slightly from the normal or half-normal solution, and these have the advantage of practical equality; that is, two approximately half-normal solutions are more convenient to work with than two which are widely different in strength. It is, however, true that some of the advantage which pertains to the use of normal solutions as regards simplicity of calculations is lost when using these approximate solutions.

The application of these general statements will be made clear in connection with the use of normal solutions in the various types of volumetric processes which follow.

I. NEUTRALIZATION METHODS

ALKALIMETRY AND ACIDIMETRY

GENERAL DISCUSSION

!Standard Acid Solutions! may be prepared from either hydrochloric, sulphuric, or oxalic acid. Hydrochloric acid has the advantage of forming soluble compounds with the alkaline earths, but its solutions cannot be boiled without danger of loss of strength; sulphuric acid solutions may be boiled without loss, but the acid forms insoluble sulphates with three of the alkaline earths; oxalic acid can be accurately weighed for the preparation of solutions, and its solutions may be boiled without loss, but it forms insoluble oxalates with three of the alkaline earths and cannot be used with certain of the indicators.

!Standard Alkali Solutions! may be prepared from sodium or potassium hydroxide, sodium carbonate, barium hydroxide, or ammonia. Of sodium and potassium hydroxide, it may be said that they can be used with all indicators, and their solutions may be boiled, but they absorb carbon dioxide readily and attack the glass of bottles, thereby losing strength; sodium carbonate may be weighed directly if its purity is assured, but the presence of carbonic acid from the carbonate is a disadvantage with many indicators; barium hydroxide solutions may be prepared which are entirely free from carbon dioxide, and such solutions immediately show by precipitation any contamination from absorption, but the hydroxide is not freely soluble in water; ammonia does not absorb carbon dioxide as readily as the caustic alkalies, but its solutions cannot be boiled nor can they be used with all indicators. The choice of a solution must depend upon the nature of the work in hand.

A !normal acid solution! should contain in one liter that quantity of the reagent which represents 1 gram of hydrogen replaceable by a base. For example, the normal solution of hydrochloric acid (HCl) should contain 36.46 grams of gaseous hydrogen chloride, since that amount furnishes the requisite 1 gram of replaceable hydrogen. On the other hand, the normal solution of sulphuric acid (H_2SO_4)

should contain only 49.03 grams, i.e., one half of its molecular weight in grams.

A !normal alkali solution! should contain sufficient alkali in a liter to replace 1 gram of hydrogen in an acid. This quantity is represented by the molecular weight in grams (40.01) of sodium hydroxide (NaOH), while a sodium carbonate solution (Na_2CO_3) should contain but one half the molecular weight in grams (i.e., 53.0 grams) in a liter of normal solution.

Half-normal or tenth-normal solutions are employed in most analyses (except in the case of the less soluble barium hydroxide). Solutions of the latter strength yield more accurate results when small percentages of acid or alkali are to be determined.

INDICATORS

It has already been pointed out that the purpose of an indicator is to mark (usually by a change of color) the point at which just enough of the titrating solution has been added to complete the chemical change which it is intended to bring about. In the neutralization processes which are employed in the measurement of alkalies (!alkalimetry!) or acids (!acidimetry!) the end-point of the reaction should, in principle, be that of complete neutrality. Expressed in terms of ionic reactions, it should be the point at which the H^{+} ions from an acid[Note 1] unite with a corresponding number of OH^{-} ions from a base to form water molecules, as in the equation

$$H^{+}, Cl^{-} + Na^{+}, OH^{-} \longrightarrow Na^{+}, Cl^{-} + (H_2O).$$

It is not usually possible to realize this condition of exact neutrality, but it is possible to approach it with sufficient exactness for analytical purposes, since substances are known which, in solution, undergo a sharp change of color as soon as even a minute excess of H^{+} or OH^{-} ions are present. Some, as will be seen, react sharply in the presence of H^{+} ions, and others with OH^{-} ions. These substances employed as indicators are usually organic compounds of complex structure and are closely allied to the dyestuffs in character.

[Note 1: A knowledge on the part of the student of the ionic theory as applied to aqueous solutions of electrolytes is assumed. A brief outline of the more important applications of the theory is given in the Appendix.]

BEHAVIOR OF ORGANIC INDICATORS

The indicators in most common use for acid and alkali titrations are methyl orange, litmus, and phenolphthalein.

In the following discussion of the principles underlying the behavior of the indicators as a class, methyl orange and phenolphthalein will be taken as types. It has just been pointed out that indicators are bodies of complicated structure. In the case of the two indicators named, the changes which they undergo have been carefully studied by Stieglitz (!J. Am. Chem. Soc.!, 25, 1112) and others, and it appears that the changes involved are of two sorts: First, a rearrangement of the atoms within the molecule, such as often occurs in organic compounds; and, second, ionic changes. The intermolecular changes cannot appropriately be discussed here, as they involve a somewhat detailed knowledge of the classification and general behavior of organic compounds; they will, therefore, be merely alluded to, and only the ionic changes followed.

Methyl orange is a representative of the group of indicators which, in aqueous solutions, behave as weak bases. The yellow color which it imparts to solutions is ascribed to the presence of the undissociated base. If an acid, such as HCl, is added to such a solution, the acid reacts with the indicator (neutralizes it) and a salt is formed, as indicated by the equation:

$(M.o.)^{+}$, OH^{-} + H^{+}, Cl^{-} —> $(M.o.)^{+}$ Cl^{-} + (H_2O).

This salt ionizes into $(M.o.)^{+}$ (using this abbreviation for the positive complex) and Cl^{-}; but simultaneously with this ionization there appears to be an internal rearrangement of the atoms which results in the production of a cation which may be designated as $(M'.o'.)^{+}$, and it is this which imparts a characteristic red color to the solution. As these changes occur in the presence of even a very small excess of acid (that is, of H^{+} ions), it serves as the desired index of their presence in the solution. If, now, an alkali, such as NaOH, is added to this reddened solution, the reverse series of changes takes place. As soon as the free acid present is neutralized, the slightest excess of sodium hydroxide, acting as a strong base, sets free the weak, little-dissociated base of the indicator, and at the moment of its formation it reverts, because of the rearrangement of the atoms, to the yellow form:

$OH^{-} + (M'.o'.)^{+} \longrightarrow [M'.o'.OH] \longrightarrow [M.o.OH]$.

Phenolphthalein, on the other hand, is a very weak, little-dissociated acid, which is colorless in neutral aqueous solution or in the presence of free H^{+} ions. When an alkali is added to such a solution, even in slight excess, the anion of the salt which has formed from the acid of the indicator undergoes a rearrangement of the atoms, and a new ion, $(Ph')^{+}$, is formed, which imparts a pink color to the solution:

$H^{+}, (Ph)^{-} + Na^{+}, OH^{-} \longrightarrow (H_2O) + Na^{+}, (Ph)^{-} \longrightarrow Na^{+}, (Ph')^{-}$

The addition of the slightest excess of an acid to this solution, on the other hand, occasions first the reversion to the colorless ion and then the setting free of the undissociated acid of the indicator:

$H^{+}, (Ph')^{-} \longrightarrow H^{+}, (Ph)^{-} \longrightarrow (HPh)$.

Of the common indicators methyl orange is the most sensitive toward alkalies and phenolphthalein toward acids; the others occupy intermediate positions. That methyl orange should be most sensitive toward alkalies is evident from the following considerations: Methyl orange is a weak base and, therefore, but little dissociated. It should, then, be formed in the undissociated condition as soon as even a slight excess of OH^{-} ions is present in the solution, and there should be a prompt change from red to yellow as outlined above. On the other hand, it should be an unsatisfactory indicator for use with weak acids (acetic acid, for example) because the salts which it forms with such acids are, like all salts of that type, hydrolyzed to a considerable extent. This hydrolytic change is illustrated by the equation:

$(M.o.)^{+} C_2H_3O_2^{-} + H^{+}, OH^{-} \longrightarrow [M.o.OH] + H^{+},$

$C_2H_3O_2^-$.

Comparison of this equation with that on page 30 will make it plain that hydrolysis is just the reverse of neutralization and must, accordingly, interfere with it. Salts of methyl orange with weak acids are so far hydrolyzed that the end-point is uncertain, and methyl orange cannot be used in the titration of such acids, while with the very weak acids, such as carbonic acid or hydrogen sulphide (hydrosulphuric acid), the salts formed with methyl orange are, in effect, completely hydrolyzed (i.e., no neutralization occurs), and methyl orange is accordingly scarcely affected by these acids. This explains its usefulness, as referred to later, for the titration of strong acids, such as hydrochloric acid, even in the presence of carbonates or sulphides in solution.

Phenolphthalein, on the other hand, should be, as it is, the best of the common indicators for use with weak acids. For, since it is itself a weak acid, it is very little dissociated, and its nearly undissociated, colorless molecules are promptly formed as soon as there is any free acid (that is, free H^+ ions) in the solution. This indicator cannot, however, be successfully used with weak bases, even ammonium hydroxide; for, since it is weak acid, the salts which it forms with weak alkalies are easily hydrolyzed, and as a consequence of this hydrolysis the change of color is not sharp. This indicator can, however, be successfully used with strong bases, because the salts which it forms with such bases are much less hydrolyzed and because the excess of OH^- ions from these bases also diminishes the hydrolytic action of water.

This indicator is affected by even so weak an acid as carbonic acid, which must be removed by boiling the solution before titration. It is the indicator most generally employed for the titration of organic acids.

In general, it may be stated that when a strong acid, such as hydrochloric, sulphuric or nitric acid, is titrated against a strong base, such as sodium hydroxide, potassium hydroxide, or barium hydroxide, any of these indicators may be used, since very little hydrolysis ensues. It has been noted above that the color change does not occur exactly at theoretical neutrality, from which it follows that no two indicators will show exactly the same end-point when acids and alkalis are brought together. It is plain, therefore, that the same indicator must be employed for both standardization and analysis, and that, if this is done, accurate results are obtainable.

The following table (Note 1) illustrates the variations in the volume of an alkali solution (tenth-normal sodium hydroxide) required to produce an alkaline end-point when run into 10 cc. of tenth-normal sulphuric acid, diluted with 50 cc. of water, using five drops of each of the different indicator solutions.

INDICATOR	N/10 H_2SO_4	N/10 NaOH SOLUTION	COLOR IN ACID SOLUTION	COLOR IN ALKALINE SOLUTION
	cc.	cc.		
Methyl orange	10	9.90	Red	Yellow
Lacmoid	10	10.00	Red	Blue
Litmus	10	10.00	Red	Blue
Rosalic acid	10	10.07	Yellow	Pink
Phenolphthalein	10	10.10	Colorless	Pink

It should also be stated that there are occasionally secondary changes, other than those outlined above, which depend upon the temperature and concentration of the solutions in which the indicators are used. These changes may influence the sensitiveness of an indicator. It is important, therefore, to take pains to use approximately the same volume of solution when standardizing that is likely to be employed in analysis; and when it is necessary, as is often the case, to titrate the solution at boiling temperature, the standardization should take place under the same conditions. It is also obvious that since some acid or alkali is required to react with the indicator itself, the amount of indicator used should be uniform and not excessive. Usually a few drops of solution will suffice.

The foregoing statements with respect to the behavior of indicators present the subject in its simplest terms. Many substances other than those named may be employed, and they have been carefully studied to determine the exact concentration of H^{+} ions at which the color change of each occurs. It is thus possible to select an indicator for a particular purpose with considerable accuracy. As data of this nature do not belong in an introductory manual, reference is made to the following papers or books in which a more extended treatment of the subject may be found:

Washburn, E.W., Principles of Physical Chemistry (McGraw-Hill Book
Co.), (Second Edition, 1921), pp. 380-387.

Prideaux, E.B.R., The Theory and Use of Indicators (Constable & Co.,
Ltd.), (1917).

Salm, E., A Study of Indicators, !Z. physik. Chem.!, 57 (1906), 471-501.

Stieglitz, J., Theories of Indicators, !J. Am. Chem. Soc.!, 25 (1903), 1112-1127.

Noyes, A.A., Quantitative Applications of the Theory of Indicators to
Volumetric Analysis, !J. Am. Chem. Soc.!, 32 (1911), 815-861.

Bjerrum, N., General Discussion, !Z. Anal. Chem.!, 66 (1917), 13-28 and 81-95.

Ostwald, W., Colloid Chemistry of Indicators, !Z. Chem. Ind. Kolloide!, 10 (1912), 132-146.

[Note 1: Glaser, !Indikatoren der Acidimetrie und Alkalimetrie!. Wiesbaden, 1901.]

PREPARATION OF INDICATOR SOLUTIONS

A !methyl orange solution! for use as an indicator is commonly made by dissolving 0.05-0.1 gram of the compound (also known as Orange III) in a few cubic centimeters of alcohol and diluting with water to 100 cc. A good grade of material should be secured. It can be successfully used for the titration of hydrochloric, nitric, sulphuric, phosphoric, and sulphurous acids, and is particularly useful in the determination of bases, such as sodium, potassium, barium, calcium, and ammonium hydroxides, and even many of the weak organic bases. It can also be used for the determination, by titration with a standard solution of a strong acid, of the salts of very weak acids, such as carbonates, sulphides, arsenites, borates, and silicates, because the weak acids which are liberated do not affect the indicator, and the reddening of the solution does not take place until an excess of the strong acid is added. It should be used in cold, not too dilute, solutions. Its sensitiveness is lessened in the presence of considerable quantities of the salts of the alkalies.

A !phenolphthalein solution! is prepared by dissolving 1 gram of the pure compound in 100 cc. of 95 per cent alcohol. This indicator is particularly valuable in the determination of weak acids, especially organic acids. It cannot be used with weak bases, even ammonia. It is affected by carbonic acid, which must, therefore, be removed by boiling when other acids are to be measured. It can be used in hot solutions. Some care is necessary to keep the volume of the solutions to be titrated approximately uniform in standardization and in analysis, and this volume should not in general exceed 125-150 cc. for the best results, since the compounds formed by the indicator undergo changes in very dilute solution which lessen its sensitiveness.

The preparation of a !solution of litmus! which is suitable for use as an indicator involves the separation from the commercial litmus of azolithmine, the true coloring principle. Soluble litmus tablets are often obtainable, but the litmus as commonly supplied to the market is mixed with calcium carbonate or sulphate and compressed into lumps. To prepare a solution, these are powdered and treated two or three times with alcohol, which dissolves out certain constituents which cause a troublesome intermediate color if not removed. The alcohol is decanted and drained off, after which the litmus is extracted with hot water until exhausted. The solution is allowed to settle for some time, the clear liquid siphoned off, concentrated to one-third its volume and acetic acid added in slight excess. It is then concentrated to a sirup, and a large excess of 95 per cent. alcohol added to it. This precipitates the blue coloring matter, which is filtered off, washed with alcohol, and finally dissolved in a small volume of water and diluted until about three drops of the solution added to 50 cc. of water just produce a distinct color. This solution must be kept in an unstoppered bottle. It should be protected from dust by a loose plug of absorbent cotton. If kept in a closed bottle it soon undergoes a reduction and loses its color, which, however, is often restored by exposure to the air.

Litmus can be employed successfully with the strong acids and bases, and also with ammonium hydroxide, although the salts of the latter influence the indicator unfavorably if present in considerable concentration. It may be employed with some of the stronger organic acids, but the use of phenolphthalein is to be preferred.

PREPARATION OF STANDARD SOLUTIONS

!Hydrochloric Acid and Sodium Hydroxide. Approximate Strength!, 0.5 N

PROCEDURE.—Measure out 40 cc. of concentrated, pure hydrochloric acid into a clean liter bottle, and dilute with distilled water to an approximate volume of 1000 cc. Shake the solution vigorously for a full minute to insure uniformity. Be sure that the bottle is not too full to permit of a thorough mixing, since lack of care at this point will be the cause of much wasted time (Note 1).

Weigh out, upon a rough balance, 23 grams of sodium hydroxide (Note 2). Dissolve the hydroxide in water in a beaker. Pour the solution into a liter bottle and dilute, as above, to approximately 1000 cc. This bottle should preferably have a rubber stopper, as the hydroxide solution attacks the glass of the ground joint of a glass stopper, and may cement the stopper to the bottle. Shake the solution as described above.

[Note 1: The original solutions are prepared of a strength greater than 0.5 N, as they are more readily diluted than strengthened if later adjustment is desired.

Too much care cannot be taken to insure perfect uniformity of solutions before standardization, and thoroughness in this respect will, as stated, often avoid much waste of time. A solution once thoroughly mixed remains uniform.]

[Note 2: Commercial sodium hydroxide is usually impure and always contains more or less carbonate; an allowance is therefore made for this impurity by placing the weight taken at 23 grams per liter. If the hydroxide is known to be pure, a lesser amount (say 21 grams) will suffice.]

COMPARISON OF ACID AND ALKALI SOLUTIONS

PROCEDURE.—Rinse a previously calibrated burette three times with the hydrochloric acid solution, using 10 cc. each time, and allowing the liquid to run out through the tip to displace all water and air from that part of the burette. Then fill the burette with the acid solution. Carry out the same procedure with a second burette, using the sodium hydroxide solution.

The acid solution may be placed in a plain or in a glass-stoppered burette as may be more convenient, but the alkaline solution should never be allowed to remain long in a glass-stoppered burette, as it tends to cement the stopper to the burette, rendering it useless. It is preferable to use a plain burette for this solution.

When the burettes are ready for use and all air bubbles displaced from the tip (see Note 2, page 17) note the exact position of the liquid in each, and record the readings in the notebook. (Consult page 188.) Run out from the burette into a beaker about 40 cc. of the acid and add two drops of a solution of methyl orange; dilute the acid to about 80 cc. and run out alkali solution from the other burette, stirring constantly, until the pink has given place to a yellow. Wash down the sides of the beaker with a little distilled water if the solution has spattered upon them, return the beaker to the acid burette, and add acid to restore the pink; continue these alternations until the point is accurately fixed at which a single drop of either solutions served to produce a distinct change of color. Select as the final end-point the appearance of the faintest pink tinge which can be recognized, or the disappearance of this tinge, leaving a pure yellow; but always titrate to the same point (Note 1). If the titration has occupied more than the three minutes required for draining the sides of the burette, the final reading may be taken immediately and recorded in the notebook.

Refill the burettes and repeat the titration. From the records of calibration already obtained, correct the burette readings and make corrections for temperature, if necessary. Obtain the ratio of the sodium hydroxide solution to that of hydrochloric acid by dividing the number of cubic centimeters of acid used by the number of cubic centimeters of alkali required for neutralization. The check results of the two titrations should not vary by more than two parts in one thousand (Note 2). If the variation in results is greater than this, refill the burettes and repeat the titration until satisfactory values are obtained. Use a new page in the notebook for each titration. Inaccurate values should not be erased or discarded. They should be retained and marked "correct" or "incorrect," as indicated by the final outcome of the titrations. This custom should be rigidly followed in all analytical work.

[Note 1: The end-point should be chosen exactly at the point of change; any darker tint is unsatisfactory, since it is impossible to carry shades of color in the memory and to duplicate them from day to day.]

[Note 2: While variation of two parts in one thousand in the values obtained by an inexperienced analyst is not excessive, the idea must be carefully avoided that this is a standard for accurate work to be !generally applied!. In many cases, after experience is gained, the allowable error is less than this proportion. In a few cases a larger variation is permissible, but these are rare and can only be recognized by an experienced analyst. It is essential that the beginner should acquire at least the degree of accuracy indicated if he is to become a successful analyst.]

STANDARDIZATION OF HYDROCHLORIC ACID

SELECTION AND PREPARATION OF STANDARD

The selection of the best substance to be used as a standard for acid solutions has been the subject of much controversy. The work of Lunge (!Ztschr. angew. Chem.! (1904), 8, 231), Ferguson (!J. Soc. Chem. Ind.! (1905), 24, 784), and others, seems to indicate that the best standard is sodium carbonate prepared from sodium bicarbonate by heating the latter at temperature between 270° and 300°C. The bicarbonate is easily prepared in a pure state, and at the temperatures named the decomposition takes place according to the equation

$2 HNaCO_3 \longrightarrow Na_2CO_3 + H_2O + CO_2$

and without loss of any carbon dioxide from the sodium carbonate, such as may occur at higher temperatures. The process is carried out as described below.

PROCEDURE.—Place in a porcelain crucible about 6 grams (roughly weighed) of the purest sodium bicarbonate obtainable. Rest the crucible upon a triangle of iron or copper wire so placed within a large crucible that there is an open air space of about three eighths of an inch between them. The larger crucible may be of iron, nickel or porcelain, as may be most convenient. Insert the bulb of a thermometer reading to 350°C. in the bicarbonate, supporting it with a clamp so that the bulb does not rest on the bottom of the crucible. Heat the outside crucible, using a rather small flame, and raise the temperature of the bicarbonate fairly rapidly to 270°C. Then regulate the heat in such a way that the temperature rises !slowly! to 300°C. in the course of a half-hour. The bicarbonate should be frequently stirred with a clean, dry, glass rod, and after stirring, should be heaped up around the bulb of the thermometer in such a way as to cover it. This will require attention during most of the heating, as the temperature should not be permitted to rise above 310°C. for any length of time. At the end of the half-hour remove the thermometer and transfer the porcelain crucible, which now contains sodium carbonate, to a desiccator. When it is cold, transfer the carbonate to a stoppered weighing tube or weighing-bottle.

STANDARDIZATION

PROCEDURE.—Clean carefully the outside of a weighing-tube, or weighing-bottle, containing the pure sodium carbonate, taking care to handle it as little as possible after wiping. Weigh the tube accurately to 0.0001 gram, and record the weight in the notebook. Hold the tube over the top of a beaker (200-300 cc.) and cautiously remove the stopper, making sure that no particles fall from it or from the tube elsewhere than in the beaker. Pour out from the tube a portion of the carbonate, replace the stopper and determine approximately how much has been removed. Continue this procedure until 1.00 to 1.10 grams has been taken from the tube. Then weigh the tube accurately and record the weight under the first weight in the notebook. The difference in the two weights is the weight of the carbonate transferred to the beaker. Proceed in the same way to transfer a second portion of the carbonate from the tube to another beaker of about the same size as the first. The beakers should be labeled and plainly marked to correspond with the entries in the notebook.

Pour over the carbonate in each beaker about 80 cc. of water, stir until solution is complete, and add two drops of methyl orange solution. Fill the burettes with the standard acid and alkali solutions, noting the initial readings of the burettes and temperature of the solutions. Run in acid from the burette, stirring and avoiding loss by effervescence, until the solution has become pink. Wash down the sides of the beaker with a !little! water from a wash-bottle, and then run in alkali from the other burette until the pink is replaced by yellow; then finish the titration as described on page 37. Note the readings of the burettes after the proper interval, and record them in the notebook. Repeat the procedure, using the second portion of sodium carbonate. Apply the necessary calibration corrections to the volumes of the solutions used, and correct for temperature if necessary.

From the data obtained, calculate the volume of the hydrochloric acid solution which is equivalent to the volume of sodium hydroxide solution used in this titration. Subtract this volume from the volume of hydrochloric acid. The difference represents the volume of acid used to react with the sodium carbonate. Divide the weight of sodium carbonate by this volume in cubic centimeters, thus obtaining the weight of sodium carbonate equivalent to each cubic centimeter of the acid.

From this weight it is possible to calculate the corresponding weight of HCl in each cubic centimeter of the acid, and in turn the relation of the acid to the normal.

If, however, it is recalled that normal solutions are equivalent to each other, it will be seen that the same result may be more readily reached by dividing the weight in grams of sodium carbonate per cubic centimeter just found by titration by the weight which would be contained in the same volume of a normal solution of sodium carbonate. A normal solution of sodium carbonate contains 53.0 grams per liter, or 0.0530 gram per cc. (see page 29). The relation of the acid solution to the normal is, therefore, calculated by dividing the weight of the carbonate to which each cubic centimeter of the acid is equivalent by 0.0530. The standardization must be repeated until the values obtained agree within, at most, two parts in one thousand.

When the standard of the acid solution has been determined, calculate, from the known ratio of the two solutions, the relation of the sodium hydroxide solution to a normal solution (Notes 1 and 2).

[Note 1: In the foregoing procedure the acid solution is standardized and the alkali solution referred to this standard by calculation. It is equally possible, if preferred, to standardize the alkali solution. The standards in a common use for this purpose are purified oxalic acid ($H_2C_2O_4 \cdot 2H_2O$), potassium acid oxalate ($KHC_2O_4 \cdot H_2O$ or KHC_2O_4), potassium tetroxalate ($KHC_2O_4 \cdot H_2C_2O_4 \cdot 2H_2O$), or potassium acid tartrate (KHC_4O_6), with the use of a suitable indicator. The oxalic acid and the oxalates should be specially prepared to insure purity, the main difficulty lying in the preservation of the water of crystallization.

It should be noted that the acid oxalate and the acid tartrate each contain one hydrogen atom replaceable by a base, while the tetroxalate contains three such atoms and the oxalic acid two. Each of the two salts first named behave, therefore, as monobasic acids, and the tetroxalate as a tribasic acid.]

[Note 2: It is also possible to standardize a hydrochloric acid solution by precipitating the chloride ions as silver chloride and weighing the precipitate, as prescribed under the analysis of sodium chloride to be described later. Sulphuric acid solutions may be standardized by precipitation of the sulphate ions as barium sulphate and weighing the ignited precipitate, but the results are not above criticism on account of the difficulty in obtaining large precipitates of barium sulphate which are uncontaminated by inclosures or are not reduced on ignition.]

DETERMINATION OF THE TOTAL ALKALINE STRENGTH OF SODA ASH

Soda ash is crude sodium carbonate. If made by the ammonia process it may contain also sodium chloride, sulphate, and hydroxide; when made by the Le Blanc process it may contain sodium sulphide, silicate, and aluminate, and other impurities. Some of these, notably the hydroxide, combine with acids and contribute to the total alkaline strength, but it is customary to calculate this strength in terms of sodium carbonate; i.e., as though no other alkali were present.

PROCEDURE.—In order to secure a sample which shall represent the average value of the ash, it is well to take at least 5 grams. As this is too large a quantity for convenient titration, an aliquot portion of the solution is measured off, representing one fifth of the entire quantity. This is accomplished as follows: Weigh out on an analytical balance two samples of soda ash of about 5 grams each into beakers of about 500 cc. capacity. (The weighings need be made to centigrams only.) Dissolve the ash in 75 cc. of water, warming gently, and filter off the insoluble residue; wash the filter by filling it at least three times with distilled water, and allowing it to drain, adding the washings to the main filtrate. Cool the filtrate to approximately the standard temperature of the laboratory, and transfer it to a 250 cc. measuring flask, washing out the beaker thoroughly. Add distilled water of laboratory temperature until the lowest point of the meniscus is level with the graduation on the neck of the flask and remove any drops of water that may be on the neck above the graduation by means of a strip of filter paper; make the solution thoroughly uniform by pouring it out into a dry beaker and back into the flask several times. Measure off 50 cc. of the solution in a measuring flask, or pipette, either of which before use should, unless they are dry on the inside, be rinsed out with at least two small portions of the soda ash solution to displace any water.

If a flask is used, fill it to the graduation with the soda ash solution and remove any liquid from the neck above the graduation with filter paper. Empty it into a beaker, and wash out the small flask, unless it is graduated for !delivery!, using small quantities of water, which are added to the liquid in the beaker. A second 50 cc. portion from the main solution should be measured off into a second beaker. Dilute the solutions in each beaker to 100 cc., add two drops of a solution of methyl orange (Note 1) and titrate for the alkali with the standard hydrochloric acid solution, using the alkali solution to complete the titration as already prescribed.

From the volumes of acid and alkali employed, corrected for burette errors and temperature changes, and the data derived from the standardization, calculate the percentage of alkali present, assuming it all to be present as sodium carbonate (Note 2).

[Note 1: The hydrochloric acid sets free carbonic acid which is unstable and breaks down into water and carbon dioxide, most of which escapes from the solution. Carbonic acid is a weak acid and, as such, does not yield a sufficient concentration of H^{+} ions to cause the indicator to change to a pink (see page 32).

The chemical changes involved may be summarized as follows:

$2H^{+}, 2Cl^{-} + 2Na^{+}, CO_{3}^{-} \longrightarrow 2Na^{+}, 2Cl^{-} + [H_{2}CO_{3}] \longrightarrow H_{2}O + CO_{2}]$

[Note 2: A determination of the alkali present as hydroxide in soda ash may be determined by precipitating the carbonate by the addition of barium chloride, removing the barium carbonate by filtration, and titrating the alkali in the filtrate.

The caustic alkali may also be determined by first using phenolphthalein as an indicator, which will show by its change from pink to colorless the point at which the caustic alkali has been neutralized and the carbonate has been converted to bicarbonate, and then adding methyl orange and completing the titration. The amount of acid necessary to change the methyl orange to pink is a measure of one half of the carbonate present. The results of the double titration furnish the data necessary for the determination of the caustic alkali and of the carbonate in the sample.]

DETERMINATION OF THE ACID STRENGTH OF OXALIC ACID

PROCEDURE.—Weigh out two portions of the acid of about 1 gram each. Dissolve these in 50 cc. of warm water. Add two drops of phenolphthalein solution, and run in alkali from the burette until the solution is pink; add acid from the other burette until the pink is just destroyed, and then add 0.3 cc. (not more) in excess. Heat the solution to boiling for three minutes. If the pink returns during the boiling, discharge it with acid and again add 0.3 cc. in excess and repeat the boiling (Note 1). If the color does not then reappear, add alkali until it does, and a !drop or two! of acid in excess and boil again for one minute (Note 2). If no color reappears during this time, complete the titration in the hot solution. The end-point should be the faintest visible shade of color (or its disappearance), as the same difficulty would exist here as with methyl orange if an attempt were made to match shades of pink.

From the corrected volume of alkali required to react with the oxalic acid, calculate the percentage of the crystallized acid ($H_2C_2O_4 \cdot 2H_2O$) in the sample (Note 3).

[Note 1: All commercial caustic soda such as that from which the standard solution was made contains some sodium carbonate. This reacts with the oxalic acid, setting free carbonic acid, which, in turn, forms sodium bicarbonate with the remaining carbonate:

$H_2CO_3 + Na_2CO_3 \longrightarrow 2HNaCO_3$.

This compound does not hydrolyze sufficiently to furnish enough OH^- ions to cause phenolphthalein to remain pink; hence, the color of the indicator is discharged in cold solutions at the point at which bicarbonate is formed. If, however, the solution is heated to boiling, the bicarbonate loses carbon dioxide and water, and reverts to sodium carbonate, which causes the indicator to become again pink:

$2HNaCO_3 \longrightarrow H_2O + CO_2 + Na_2CO_3$.

By adding successive portions of hydrochloric acid and boiling, the carbonate is ultimately all brought into reaction.

The student should make sure that the difference in behavior of the two indicators, methyl orange and phenolphthalein, is understood.]

[Note 2: Hydrochloric acid is volatilized from aqueous solutions, except such as are very dilute. If the directions in the procedure are strictly followed, no loss of acid need be feared, but the amount added in excess should not be greater than 0.3-0.4 cc.]

[Note 3: Attention has already been called to the fact that the color changes in the different indicators occur at varying concentrations of H^{+} or OH^{-} ions. They do not indicate exact theoretical neutrality, but a particular indicator always shows its color change at a particular concentration of H^{+} or OH^{-} ions. The results of titration with a given indicator are, therefore, comparable. As a matter of fact, a small error is involved in the procedure as outlined above. The comparison of the acid and alkali solutions was made, using methyl orange as an indicator, while the titration of the oxalic acid is made with the use of phenolphthalein. For our present purposes the small error may be neglected but, if time permits, the student is recommended to standardize the alkali solution against one of the substances named in Note 1, page 41, and also to ascertain the comparative value of the acid and alkali solutions, using phenolphthalein as indicator throughout, and conducting the titrations as described above. This will insure complete accuracy.]

II. OXIDATION PROCESSES

GENERAL DISCUSSION

In the oxidation processes of volumetric analysis standard solutions of oxidizing agents and of reducing agents take the place of the acid and alkali solutions of the neutralization processes already studied. Just as an acid solution was the principal reagent in alkalimetry, and the alkali solution used only to make certain of the end-point, the solution of the oxidizing agent is the principal reagent for the titration of substances exerting a reducing action. It is, in general, true that oxidizable substances are determined by !direct! titration, while oxidizing substances are determined by !indirect! titration.

The important oxidizing agents employed in volumetric solutions are potassium bichromate, potassium permangenate, potassium ferricyanide, iodine, ferric chloride, and sodium hypochlorite.

The important reducing agents which are used in the form of standard solutions are ferrous sulphate (or ferrous ammonium sulphate), oxalic acid, sodium thiosulphate, stannous chloride, arsenious acid, and potassium cyanide. Other reducing agents, as sulphurous acid, sulphureted hydrogen, and zinc (nascent hydrogen), may take part in the processes, but not as standard solutions.

The most important combinations among the foregoing are: Potassium bichromate and ferrous salts; potassium permanganate and ferrous salts; potassium permanganate and oxalic acid, or its derivatives; iodine and sodium thiosulphate; hypochlorites and arsenious acid.

BICHROMATE PROCESS FOR THE DETERMINATION OF IRON

Ferrous salts may be promptly and completely oxidized to ferric salts, even in cold solution, by the addition of potassium bichromate, provided sufficient acid is present to hold in solution the ferric and chromic compounds which are formed.

The acid may be either hydrochloric or sulphuric, but the former is usually preferred, since it is by far the best solvent for iron and its compounds. The reaction in the presence of hydrochloric acid is as follows:

$6FeCl_2 + K_2Cr_2O_7 + 14HCl \longrightarrow 6FeCl_3 + 2CrCl_3 + 2KCl + 7H_2O$.

NORMAL SOLUTIONS OF OXIDIZING OR REDUCING AGENTS

It will be recalled that the system of normal solutions is based upon the equivalence of the reagents which they contain to 8 grams of oxygen or 1 gram of hydrogen. A normal solution of an oxidizing agent should, therefore, contain that amount per liter which is equivalent in oxidizing power to 8 grams of oxygen; a normal reducing solution must be equivalent in reducing power to 1 gram of hydrogen. In order to determine what the amount per liter will be it is necessary to know how the reagents enter into reaction. The two solutions to be employed in the process under consideration are those of potassium bichromate and ferrous sulphate. The reaction between them, in the presence of an excess of sulphuric acid, may be expressed as follows:

$6FeSO_4 + K_2Cr_2O_7 + 7H_2SO_4 \longrightarrow 3Fe_2(SO_4)_3 + K_2SO_4 + Cr_2(SO_4)_3 + 7H_2O$.

If the compounds of iron and chromium, with which alone we are now concerned, be written in such a way as to show the oxides of these elements in each, they would appear as follows: On the left-hand side of the equation 6(FeO.SO$_3$) and K$_2$O.2CrO$_3$; on the right-hand side, 3(Fe$_2$O$_3$.3SO$_3$) and Cr$_2$O$_3$.3SO$_3$. A careful inspection shows that there are three less oxygen atoms associated with chromium atoms on the right-hand side of the equation than on the left-hand, but there are three more oxygen atoms associated with iron atoms on the right than on the left. In other words, a molecule of potassium bichromate has given up three atoms of oxygen for oxidation purposes; i.e., a molecular weight in grams of the bichromate (294.2) will furnish 3 X 16 or 48 grams of oxygen for oxidation purposes. As this 48 grams is six times 8 grams, the basis of the system, the normal solution of potassium bichromate should contain per liter one sixth of 294.2 grams or 49.03 grams.

A further inspection of the dissected compounds above shows that six molecules of FeO.SO$_3$ were required to react with the three atoms of oxygen from the bichromate. From the two equations

3H$_2$ + 3O —> 3H$_2$O 6(FeO.SO$_3$) + 3O —> 3(Fe$_2$O$_3$.3SO$_3$)

it is plain that one molecule of ferrous sulphate is equivalent to one atom of hydrogen in reducing power; therefore one molecular weight in grams of ferrous sulphate (151.9) is equivalent to 1 gram of hydrogen. Since the ferrous sulphate crystalline form has the formula FeSO$_4$.7H$_2$O, a normal reducing solution of this crystalline salt should contain 277.9 grams per liter.

PREPARATION OF SOLUTIONS

!Approximate Strength 0.1 N!

It is possible to purify commercial potassium bichromate by recrystallization from hot water. It must then be dried and cautiously heated to fusion to expel the last traces of moisture, but not sufficiently high to expel any oxygen. The pure salt thus prepared, may be weighed out directly, dissolved, and the solution diluted in a graduated flask to a definite volume. In this case no standardization is made, as the normal value can be calculated directly. It is, however, more generally customary to standardize a solution of the commercial salt by comparison with some substance of definite composition, as described below.

PROCEDURE.—Pulverize about 5 grams of potassium bichromate of good quality. Dissolve the bichromate in distilled water, transfer the solution to a liter bottle, and dilute to approximately 1000 cc. Shake thoroughly until the solution is uniform.

To prepare the solution of the reducing agent, pulverize about 28 grams of ferrous sulphate ($FeSO_4 \cdot 7H_2O$) or about 40 grams of ferrous ammonium sulphate ($FeSO_4 \cdot (NH_4)_2SO_4 \cdot 6H_2O$) and dissolve in distilled water containing 5 cc. of concentrated sulphuric acid. Transfer the solution to a liter bottle, add 5 cc. concentrated sulphuric acid, make up to about 1000 cc. and shake vigorously to insure uniformity.

INDICATOR SOLUTION

No indicator is known which, like methyl orange, can be used within the solution, to show when the oxidation process is complete. Instead, an outside indicator solution is employed to which drops of the titrated solution are transferred for testing. The reagent used is potassium ferricyanide, which produces a blue precipitate (or color) with ferrous compounds as long as there are unoxidized ferrous ions in the titrated solution. Drops of the indicator solution are placed upon a glazed porcelain tile, or upon white cardboard which has been coated with paraffin to render it waterproof, and drops of the titrated solution are transferred to the indicator on the end of a stirring rod. When the oxidation is nearly completed only very small amounts of the ferrous compounds remain unoxidized and the reaction with the indicator is no longer instantaneous. It is necessary to allow a brief time to elapse before determining that no blue color is formed. Thirty seconds is a sufficient interval, and should be adopted throughout the analytical procedure. If left too long, the combined effect of light and dust from the air will cause a reduction of the ferric compounds already formed and a resultant blue will appear which misleads the observer with respect to the true end-point.

The indicator solution must be highly diluted, otherwise its own color interferes with accurate observation. Prepare a fresh solution, as needed each day, by dissolving a crystal of potassium ferricyanide about the size of a pin's head in 25 cc. of distilled water. The salt should be carefully tested with ferric chloride for the presence of ferrocyanides, which give a blue color with ferric salts.

In case of need, the ferricyanide can be purified by adding to its solution a little bromine water and recrystallizing the compound.

COMPARISON OF OXIDIZING AND REDUCING SOLUTIONS

PROCEDURE.—Fill one burette with each of the solutions, observing the general procedure with respect to cleaning and rinsing already prescribed. The bichromate solution is preferably to be placed in a glass-stoppered burette.

Run out from a burette into a beaker of about 300 cc. capacity nearly 40 cc. of the ferrous solution, add 15 cc. of dilute hydrochloric acid (sp. gr. 1.12) and 150 cc. of water and run in the bichromate solution from another burette. Since both solutions are approximately tenth-normal, 35 cc. of the bichromate solution may be added without testing. Test at that point by removing a very small drop of the iron solution on the end of a stirring rod, mixing it with a drop of indicator on the tile (Note 1). If a blue precipitate appears at once, 0.5 cc. of the bichromate solution may be added before testing again. The stirring rod which has touched the indicator should be dipped in distilled water before returning it to the iron solution. As soon as the blue appears to be less intense, add the bichromate solution in small portions, finally a single drop at a time, until the point is reached at which no blue color appears after the lapse of thirty seconds from the time of mixing solution and indicator. At the close of the titration a large drop of the iron solution should be taken for the test. To determine the end-point beyond any question, as soon as the thirty seconds have elapsed remove another drop of the solution of the same size as that last taken and mix it with the indicator, placing it beside the last previous test. If this last previous test shows a blue tint in comparison with the fresh mixture, the end-point has not been reached; if no difference can be noted the reaction is complete. Should the end-point be overstepped, a little more of the ferrous solution may be added and the end-point definitely fixed.

From the volumes of the solutions used, after applying corrections for burette readings, and, if need be, for the temperature of solutions, calculate the value of the ferrous solution in terms of the oxidizing solution.

[Note 1: The accuracy of the work may be much impaired by the removal of unnecessarily large quantities of solution for the tests. At the beginning of the titration, while much ferrous iron is still present, the end of the stirring rod need only be moist with the solution; but at the close of the titration drops of considerable size may properly be taken for the final tests. The stirring rod should be washed to prevent transfer of indicator to the main solution. This cautious removal of solution does not seriously affect the accuracy of the determination, as it will be noted that the volume of the titrated solution is about 200 cc. and the portions removed are very small. Moreover, if the procedure is followed as prescribed, the concentration of unoxidized iron decreases very rapidly as the titration is carried out so that when the final tests are made, though large drops may be taken, the amount of ferrous iron is not sufficient to produce any appreciable error in results.

If the end-point is determined as prescribed, it can be as accurately fixed as that of other methods; and if a ferrous solution is at hand, the titration need consume hardly more time than that of the permanganate process to be described later on.]

STANDARDIZATION OF POTASSIUM BICHROMATE SOLUTIONS

!Selection of a Standard!

A substance which will serve satisfactorily as a standard for oxidizing solutions must possess certain specific properties: It must be of accurately known composition and definite in its behavior as a reducing agent, and it must be permanent against oxidation in the air, at least for considerable periods. Such standards may take the form of pure crystalline salts, such as ferrous ammonium sulphate, or may be in the form of iron wire or an iron ore of known iron content. It is not necessary that the standard should be of 100 per cent purity, provided the content of the active reducing agent is known and no interfering substances are present.

The two substances most commonly used as standards for a bichromate solution are ferrous ammonium sulphate and iron wire. A standard wire is to be purchased in the market which answers the purpose well, and its iron content may be determined for each lot purchased by a number of gravimetric determinations. It may best be preserved in jars containing calcium chloride, but this must not be allowed to come into contact with the wire. It should, however, even then be examined carefully for rust before use.

If pure ferrous ammonium sulphate is used as the standard, clear crystals only should be selected. It is perhaps even better to determine by gravimetric methods once for all the iron content of a large commercial sample which has been ground and well mixed. This salt is permanent over long periods if kept in stoppered containers.

STANDARDIZATION

PROCEDURE.—Weigh out two portions of iron wire of about 0.24-0.26 gram each, examining the wire carefully for rust. It should be handled and wiped with filter paper (not touched by the fingers), should be weighed on a watch-glass, and be bent in such a way as not to interfere with the movement of the balance.

Place 30 cc. of hydrochloric acid (sp. gr. 1.12) in each of two 300 cc. Erlenmeyer flasks, cover them with watch-glasses, and bring the acid just to boiling. Remove them from the flame and drop in the portions of wire, taking great care to avoid loss of liquid during solution. Boil for two or three minutes, keeping the flasks covered (Note 1), then wash the sides of the flasks and the watch-glass with a little water and add stannous chloride solution to the hot liquid !from a dropper! until the solution is colorless, but avoid more than a drop or two in excess (Note 2). Dilute with 150 cc. of water and cool !completely!. When cold, add rapidly about 30 cc. of mercuric chloride solution. Allow the solutions to stand about three minutes and then titrate without further delay (Note 3), add about 35 cc. of the standard solution at once and finish the titration as prescribed above, making use of the ferrous solution if the end-point should be passed.

From the corrected volumes of the bichromate solution required to oxidize the iron actually know to be present in the wire, calculate the relation of the standard solution to the normal.

Repeat the standardization until the results are concordant within at least two parts in one thousand.

[Note 1: The hydrochloric acid is added to the ferrous solution to insure the presence of at least sufficient free acid for the titration, as required by the equation on page 48.

The solution of the wire in hot acid and the short boiling insure the removal of compounds of hydrogen and carbon which are formed from the small amount of carbon in the iron. These might be acted upon by the bichromate if not expelled.]

[Note 2: It is plain that all the iron must be reduced to the ferrous condition before the titration begins, as some oxidation may have occurred from the oxygen of the air during solution. It is also evident that any excess of the agent used to reduce the iron must be removed; otherwise it will react with the bichromate added later.

The reagents available for the reduction of iron are stannous chloride, sulphurous acid, sulphureted hydrogen, and zinc; of these stannous chloride acts most readily, the completion of the reaction is most easily noted, and the excess of the reagent is most readily removed. The latter object is accomplished by oxidation to stannic chloride by means of mercuric chloride added in excess, as the mercuric salts have no effect upon ferrous iron or the bichromate. The reactions involved are:

$2FeCl_3 + SnCl_2 \rightarrow 2FeCl_2 + SnCl_4$
$SnCl_2 + 2HgCl_2 \rightarrow SnCl_4 + 2HgCl$

The mercurous chloride is precipitated.

It is essential that the solution should be cold and that the stannous chloride should not be present in great excess, otherwise a secondary reaction takes place, resulting in the reduction of the mercurous chloride to metallic mercury:

$SnCl_2 + 2HgCl \longrightarrow SnCl_4 + 2Hg$.

The occurrence of this secondary reaction is indicated by the darkening of the precipitate; and, since potassium bichromate oxidizes this mercury slowly, solutions in which it has been precipitated are worthless as iron determinations.]

[Note 3: The solution should be allowed to stand about three minutes after the addition of mercuric chloride to permit the complete deposition of mercurous chloride. It should then be titrated without delay to avoid possible reoxidation of the iron by the oxygen of the air.]

DETERMINATION OF IRON IN LIMONITE

PROCEDURE.—Grind the mineral (Note 1) to a fine powder. Weigh out accurately two portions of about 0.5 gram (Note 2) into porcelain crucibles; heat these crucibles to dull redness for ten minutes, allow them to cool, and place them, with their contents, in beakers containing 30 cc. of dilute hydrochloric acid (sp. gr. 1.12). Heat at a temperature just below boiling until the undissolved residue is white or until solvent action has ceased. If the residue is white, or known to be free from iron, it may be neglected and need not be removed by filtration. If a dark residue remains, collect it on a filter, wash free from hydrochloric acid, and ignite the filter in a platinum crucible (Note 3). Mix the ash with five times its weight of sodium carbonate and heat to fusion; cool, and disintegrate the fused mass with boiling water in the crucible. Unite this solution and precipitate (if any) with the acid solution, taking care to avoid loss by effervescence. Wash out the crucible, heat the acid solution to boiling, add stannous chloride solution until it is colorless, avoiding a large excess (Note 4); cool, and when !cold!, add 40 cc. of mercuric chloride solution, dilute to 200 cc., and proceed with the titration as already described.

From the standardization data already obtained, and the known weight of the sample, calculate the percentage of iron (Fe) in the limonite.

[Note 1: Limonite is selected as a representative of iron ores in general. It is a native, hydrated oxide of iron. It frequently occurs in or near peat beds and contains more or less organic matter which, if brought into solution, would be acted upon by the potassium bichromate. This organic matter is destroyed by roasting. Since a high temperature tends to lessen the solubility of ferric oxide, the heat should not be raised above low redness.]

[Note 2: It is sometimes advantageous to dissolve a large portion—say 5 grams—and to take one tenth of it for titration. The sample will then represent more closely the average value of the ore.]

[Note 3: A platinum crucible may be used for the roasting of the limonite and must be used for the fusion of the residue. When used, it must not be allowed to remain in the acid solution of ferric chloride for any length of time, since the platinum is attacked and dissolved, and the platinic chloride is later reduced by the stannous chloride, and in the reduced condition reacts with the bichromate, thus introducing an error. It should also be noted that copper and antimony interfere with the determination of iron by the bichromate process.]

[Note 4: The quantity of stannous chloride required for the reduction of the iron in the limonite will be much larger than that added to the solution of iron wire, in which the iron was mainly already in the ferrous condition. It should, however, be added from a dropper to avoid an unnecessary excess.]

DETERMINATION OF CHROMIUM IN CHROME IRON ORE

PROCEDURE.—Grind the chrome iron ore (Note 1) in an agate mortar until no grit is perceptible under the pestle. Weigh out two portions of 0.5 gram each into iron crucibles which have been scoured inside until bright (Note 2). Weigh out on a watch-glass (Note 3), using the rough balances, 5 grams of dry sodium peroxide for each portion, and pour about three quarters of the peroxide upon the ore. Mix ore and flux by thorough stirring with a dry glass rod. Then cover the mixture with the remainder of the peroxide. Place the crucible on a triangle and raise the temperature !slowly! to the melting point of the flux, using a low flame, and holding the lamp in the hand (Note 4). Maintain the fusion for five minutes, and stir constantly with a stout iron wire, but do not raise the temperature above moderate redness (Notes 5 and 6).

Allow the crucible to cool until it can be comfortably handled (Note 7) and then place it in a 300 cc. beaker, and cover it with distilled water (Note 8). The beaker must be carefully covered to avoid loss during the disintegration of the fused mass. When the evolution of gas ceases, rinse off and remove the crucible; then heat the solution !while still alkaline! to boiling for fifteen minutes. Allow the liquid to cool for a few minutes; then acidify with dilute sulphuric acid (1:5), adding 10 cc. in excess of the amount necessary to dissolve the ferric hydroxide (Note 9). Dilute to 200 cc., cool, add from a burette an excess of a standard ferrous solution, and titrate for the excess with a standard solution of potassium bichromate, using the outside indicator (Note 10).

From the corrected volumes of the two standard solutions, and their relations to normal solutions, calculate the percentage of chromium in the ore.

[Note 1: Chrome iron ore is essentially a ferrous chromite, or combination of FeO and Cr_2O_3. It must be reduced to a state of fine subdivision to ensure a prompt reaction with the flux.]

[Note 2: The scouring of the iron crucible is rendered much easier if it is first heated to bright redness and plunged into cold water. In this process oily matter is burned off and adhering scale is caused to chip off when the hot crucible contracts rapidly in the cold water.]

[Note 3: Sodium peroxide must be kept off of balance pans and should not be weighed out on paper, as is the usual practice in the rough weighing of chemicals. If paper to which the peroxide is adhering is exposed to moist air it is likely to take fire as a result of the absorption of moisture, and consequent evolution of heat and liberation of oxygen.]

[Note 4: The lamp should never be allowed to remain under the crucible, as this will raise the temperature to a point at which the crucible itself is rapidly attacked by the flux and burned through.]

[Note 5: The sodium peroxide acts as both a flux and an oxidizing agent. The chromic oxide is dissolved by the flux and oxidized to chromic anhydride (CrO_3) which combines with the alkali to form sodium chromate. The iron is oxidized to ferric oxide.]

[Note 6: The sodium peroxide cannot be used in porcelain, platinum, or silver crucibles. It attacks iron and nickel as well; but crucibles made from these metals may be used if care is exercised to keep the temperature as low as possible. Preference is here given to iron crucibles, because the resulting ferric hydroxide is more readily brought into solution than the nickelic oxide from a nickel crucible. The peroxide must be dry, and must be protected from any admixture of dust, paper, or of organic matter of any kind, otherwise explosions may ensue.]

[Note 7: When an iron crucible is employed it is desirable to allow the fusion to become nearly cold before it is placed in water, otherwise scales of magnetic iron oxide may separate from the crucible, which by slowly dissolving in acid form ferrous sulphate, which reduces the chromate.]

[Note 8: Upon treatment with water the chromate passes into solution, the ferric hydroxide remains undissolved, and the excess of peroxide is decomposed with the evolution of oxygen. The subsequent boiling insures the complete decomposition of the peroxide. Unless this is complete, hydrogen peroxide is formed when the solution is acidified, and this reacts with the bichromate, reducing it and introducing a serious error.]

[Note 9: The addition of the sulphuric acid converts the sodium chromate to bichromate, which behaves exactly like potassium bichromate in acid solution.]

[Note 10: If a standard solution of a ferrous salt is not at hand, a weight of iron wire somewhat in excess of the amount which would be required if the chromite were pure $FeO.Cr_2O_3$ may be weighed out and dissolved in sulphuric acid; after reduction of all the iron by stannous chloride and the addition of mercuric chloride, this solution may be poured into the chromate solution and the excess of iron determined by titration with standard bichromate solution.]

PERMANGANATE PROCESS FOR THE DETERMINATION OF IRON

Potassium permanganate oxidizes ferrous salts in cold, acid solution promptly and completely to the ferric condition, while in hot acid solution it also enters into a definite reaction with oxalic acid, by which the latter is oxidized to carbon dioxide and water.

The reactions involved are these:

$10FeSO_4 + 2KMnO_4 + 8H_2S_4 \longrightarrow 5Fe_2(SO_4)_3 + K_2SO_4 + 2MnSO_4 + 8H_2O$

$5C_2H_2O_4(2H_2O) + 2KMnO_4 + 3H_2SO_4 \longrightarrow K_2SO_4 + 2MnSO_4 + 10CO_2 + 1\ H_2O$.

These are the fundamental reactions upon which the extensive use of potassium permanganate depends; but besides iron and oxalic acid the permanganate enters into reaction with antimony, tin, copper, mercury, and manganese (the latter only in neutral solution), by which these metals are changed from a lower to a higher state of oxidation; and it also reacts with sulphurous acid, sulphureted hydrogen, nitrous acid, ferrocyanides, and most soluble organic bodies. It should be noted, however, that very few of these organic compounds react quantitatively with the permanganate, as is the case with oxalic acid and the oxalates.

Potassium permanganate is acted upon by hydrochloric acid; the action is rapid in hot or concentrated solution (particularly in the presence of iron salts, which appear to act as catalyzers, increasing the velocity of the reaction), but slow in cold, dilute solutions. However, the greater solubility of iron compounds in hydrochloric acid makes it desirable to use this acid as a solvent, and experiments made with this end in view have shown that in cold, dilute hydrochloric acid solution, to which considerable quantities of manganous sulphate and an excess of phosphoric acid have been added, it is possible to obtain satisfactory results.

It is also possible to replace the hydrochloric acid by evaporating the solutions with an excess of sulphuric acid until the latter fumes. This procedure is somewhat more time-consuming, but the end-point of the permanganate titration is more permanent. Both procedures are described below.

Potassium permanganate has an intense coloring power, and since the solution resulting from the oxidation of the iron and the reduction of the permanganate is colorless, the latter becomes its own indicator. The slightest excess is indicated with great accuracy by the pink color of the solution.

PREPARATION OF A STANDARD SOLUTION

!Approximate Strength 0.1 N!

A study of the reactions given above which represent the oxidation of ferrous compounds by potassium permanganate, shows that there are 2 molecules of $KMnO_4$ and 10 molecules of $FeSO_4$ on the left-hand side, and 2 molecules of $MnSO_4$ and 5 molecules of $Fe_2(SO_4)_5$ on the right-hand side. Considering only these compounds, and writing the formulas in such a way as to show the oxides of the elements in each, the equation becomes:

$K_2O.Mn_2O_7 + 10(FeO.SO_3) \rightarrow K_2O.SO_3 + 2(MnO.SO_3) + 5(Fe_2O_3.3SO_3)$.

From this it appears that two molecules of $KMnO_4$ (or 316.0 grams) have given up five atoms (or 80 grams) of oxygen to oxidize the ferrous compound. Since 8 grams of oxygen is the basis of normal oxidizing solutions and 80 grams of oxygen are supplied by 316.0 grams of $KMnO_4$, the normal solution of the permanganate should contain, per liter, 316.0/10 grams, or 31.60 grams (Note 1).

The preparation of an approximately tenth-normal solution of the reagent may be carried out as follows:

PROCEDURE.—Dissolve about 3.25 grams of potassium permanganate crystals in approximately 1000 cc. of distilled water in a large beaker, or casserole. Heat slowly and when the crystals have dissolved, boil the solution for 10-15 minutes. Cover the solution with a watch-glass; allow it to stand until cool, or preferably over night. Filter the solution through a layer of asbestos. Transfer the filtrate to a liter bottle and mix thoroughly (Note 2).

[Note 1: The reactions given on page 61 are those which take place in the presence of an excess of acid. In neutral solutions the reduction of the permanganate is less complete, and, under these conditions, two gram-molecular weights of $KMnO_4$ will furnish only 48 grams of oxygen. A normal solution for use under these conditions should, therefore, contain 316.0/6 grams, or 52.66 grams.]

[Note 2: Potassium permanganate solutions are not usually stable for long periods, and change more rapidly when first prepared than after standing some days. This change is probably caused by interaction with the organic matter contained in all distilled water, except that redistilled from an alkaline permanganate solution. The solutions should be protected from light and heat as far as possible, since both induce decomposition with a deposition of manganese dioxide, and it has been shown that decomposition proceeds with considerable rapidity, with the evolution of oxygen, after the dioxide has begun to form. As commercial samples of the permanganate are likely to be contaminated by the dioxide, it is advisable to boil and filter solutions through asbestos before standardization, as prescribed above. Such solutions are relatively stable.]

COMPARISON OF PERMANGANATE AND FERROUS SOLUTIONS

PROCEDURE.—Fill a glass-stoppered burette with the permanganate solution, observing the usual precautions, and fill a second burette with the ferrous sulphate solution prepared for use with the potassium bichromate. The permanganate solution cannot be used in burettes with rubber tips, as a reduction takes place upon contact with the rubber. The solution has so deep a color that the lower line of the meniscus cannot be detected; readings must therefore be made from the upper edge. Run out into a beaker about 40 cc. of the ferrous solution, dilute to about 100 cc., add 10 cc. of dilute sulphuric acid, and run in the permanganate solution to a slight permanent pink. Repeat, until the ratio of the two solutions is satisfactorily established.

STANDARDIZATION OF A POTASSIUM PERMANGANATE SOLUTION

!Selection of a Standard!

Commercial potassium permanganate is rarely sufficiently pure to admit of its direct weighing as a standard. On this account, and because of the uncertainties as to the permanence of its solutions, it is advisable to standardize them against substances of known value. Those in most common use are iron wire, ferrous ammonium sulphate, sodium oxalate, oxalic acid, and some other derivatives of oxalic acid. With the exception of sodium oxalate, these all contain water of crystallization which may be lost on standing. They should, therefore, be freshly prepared, and with great care. At present, sodium oxalate is considered to be one of the most satisfactory standards.

!Method A!

!Iron Standards!

The standardization processes employed when iron or its compounds are selected as standards differ from those applicable in connection with oxalate standards. The procedure which immediately follows is that in use with iron standards.

As in the case of the bichromate process, it is necessary to reduce the iron completely to the ferrous condition before titration. The reducing agents available are zinc, sulphurous acid, or sulphureted hydrogen. Stannous chloride may also be used when the titration is made in the presence of hydrochloric acid. Since the excess of both the gaseous reducing agents can only be expelled by boiling, with consequent uncertainty regarding both the removal of the excess and the reoxidation of the iron, zinc or stannous chlorides are the most satisfactory agents. For prompt and complete reduction it is essential that the iron solution should be brought into ultimate contact with the zinc. This is brought about by the use of a modified Jones reductor, as shown in Figure 1. This reductor is a standard apparatus and is used in other quantitative processes.

[Illustration: Fig. 1]

The tube A has an inside diameter of 18 mm. and is 300 mm. long; the small tube has an inside diameter of 6 mm. and extends 100 mm. below the stopcock. At the base of the tube A are placed some pieces of broken glass or porcelain, covered by a plug of glass wool about 8 mm. thick, and upon this is placed a thin layer of asbestos, such as is used for Gooch filters, 1 mm. thick. The tube is then filled with the amalgamated zinc (Note 1) to within 50 mm. of the top, and on the zinc is placed a plug of glass wool. If the top of the tube is not already shaped like the mouth of a thistle-tube (B), a 60 mm. funnel is fitted into the tube with a rubber stopper and the reductor is connected with a suction bottle, F. The bottle D is a safety bottle to prevent contamination of the solution by water from the pump. After preparation for use, or when left standing, the tube A should be filled with water, to prevent clogging of the zinc.

[Note 1: The use of fine zinc in the reductor is not necessary and tends to clog the tube. Particles which will pass a 10-mesh sieve, but are retained by one of 20 meshes to the inch, are most satisfactory. The zinc can be amalgamated by stirring or shaking it in a mixture of 25 cc. of normal mercuric chloride solution, 25 cc. of hydrochloric acid (sp. gr. 1.12) and 250 cc. of water for two minutes. The solution should then be poured off and the zinc thoroughly washed. It is then ready for bottling and preservation under water. A small quantity of glass wool is placed in the neck of the funnel to hold back foreign material when the reductor is in use.]

STANDARDIZATION

PROCEDURE.—Weigh out into Erlenmeyer flasks two portions of iron wire of about 0.25 gram each. Dissolve these in hot dilute sulphuric acid (5 cc. of concentrated acid and 100 cc. of water), using a covered flask to avoid loss by spattering. Boil the solution for two or three minutes after the iron has dissolved to remove any volatile hydrocarbons. Meanwhile prepare the reductor for use as follows: Connect the vacuum bottle with the suction pump and pour into the funnel at the top warm, dilute sulphuric acid, prepared by adding 5 cc. of concentrated sulphuric acid to 100 cc. of distilled water. See that the stopcock (C) is open far enough to allow the acid to run through slowly. Continue to pour in acid until 200 cc. have passed through, then close the stopcock !while a small quantity of liquid is still left in the funnel!. Discard the filtrate, and again pass through 100 cc. of the warm, dilute acid. Test this with the permanganate solution. A single drop should color it permanently; if it does not, repeat the washing, until assured that the zinc is not contaminated with appreciable quantities of reducing substances. Be sure that no air enters the reductor (Note 1).

Pour the iron solution while hot (but not boiling) through the reductor at a rate not exceeding 50 cc. per minute (Notes 2 and 3). Wash out the beaker with dilute sulphuric acid, and follow the iron solution without interruption with 175 cc. of the warm acid and finally with 75 cc. of distilled water, leaving the funnel partially filled. Remove the filter bottle and cool the solution quickly under the water tap (Note 4), avoiding unnecessary exposure to the oxygen of the air. Add 10 cc. of dilute sulphuric acid and titrate to a faint pink with the permanganate solution, adding it directly to the contents of the vacuum flask. Should the end-point be overstepped, the ferrous sulphate solution may be added.

From the volume of the solution required to oxidize the iron in the wire, calculate the relation to the normal of the permanganate solution. The duplicate results should be concordant within two parts in one thousand.

[Note 1: The funnel of the reductor must never be allowed to empty. If it is left partially filled with water the reductor is ready for subsequent use after a very little washing; but a preliminary test is always necessary to safeguard against error.

If more than a small drop of permanganate solution is required to color 100 cc. of the dilute acid after the reductor is well washed, an allowance must be made for the iron in the zinc. !Great care! must be used to prevent the access of air to the reductor after it has been washed out ready for use. If air enters, hydrogen peroxide forms, which reacts with the permanganate, and the results are worthless.]

[Note 2: The iron is reduced to the ferrous condition by contact with the zinc. The active agent may be considered to be !nascent! hydrogen, and it must be borne in mind that the visible bubbles are produced by molecular hydrogen, which is without appreciable effect upon ferric iron.

The rate at which the iron solution passes through the zinc should not exceed that prescribed, but the rate may be increased somewhat when the wash-water is added. It is well to allow the iron solution to run nearly, but not entirely, out of the funnel before the wash-water is added. If it is necessary to interrupt the process, the complete emptying of the funnel can always be avoided by closing the stopcock.

It is also possible to reduce the iron by treatment with zinc in a flask from which air is excluded. The zinc must be present in excess of the quantity necessary to reduce the iron and is finally completely dissolved. This method is, however, less convenient and more tedious than the use of the reductor.]

[Note 3: The dilute sulphuric acid for washing must be warmed ready for use before the reduction of the iron begins, and it is of the first importance that the volume of acid and of wash-water should be measured, and the volume used should always be the same in the standardizations and all subsequent analyses.]

[Note 4: The end-point is more permanent in cold than hot solutions, possibly because of a slight action of the permanganate upon the manganous sulphate formed during titration. If the solution turns brown, it is an evidence of insufficient acid, and more should be immediately added. The results are likely to be less accurate in this case, however, as a consequence of secondary reactions between the ferrous iron and the manganese dioxide thrown down. It is wiser to discard such results and repeat the process.]

[Note 5: The potassium permanganate may, of course, be diluted and brought to an exactly 0.1 N solution from the data here obtained. The percentage of iron in the iron wire must be taken into account in all calculations.]

!Method B!

!Oxalate Standards!

PROCEDURE.—Weigh out two portions of pure sodium oxalate of 0.25-0.3 gram each into beakers of about 600 cc. capacity. Add about 400 cc. of boiling water and 20 cc. of manganous sulphate solution (Note 1). When the solution of the oxalate is complete, heat the liquid, if necessary, until near its boiling point (70-90°C.) and run in the standard permanganate solution drop by drop from a burette, stirring constantly until an end-point is reached (Note 2). Make a blank test with 20 cc. of manganous sulphate solution and a volume of distilled water equal to that of the titrated solution to determine the volume of the permanganate solution required to produce a very slight pink. Deduct this volume from the amount of permanganate solution used in the titration.

From the data obtained, calculate the relation of the permanganate solution to the normal. The reaction involved is:

$5Na_2C_2O_4 + 2KMnO_4 + 8H_2SO_4 \longrightarrow 5Na_2SO_4 + K_2SO_4 + 2MnSO_4 + 10CO_2 + 8H_2O$

[Note 1: The manganous sulphate titrating solution is made by dissolving 20 grams of $MnSO_4$ in 200 cubic centimeters of water and adding 40 cc. of concentrated sulphuric acid (sp. gr. 1.84) and 40 cc. or phosphoric acid (85%).]

[Note 2: The reaction between oxalates and permanganates takes place quantitatively only in hot acid solutions. The temperatures must not fall below 70°C.]

DETERMINATION OF IRON IN LIMONITE

!Method A!

The procedures, as here prescribed, are applicable to iron ores in general, provided these ores contain no constituents which are reduced by zinc or stannous chloride and reoxidized by permanganates. Many iron ores contain titanium, and this element among others does interfere with the determination of iron by the process described. If, however, the solutions of such ores are treated with sulphureted hydrogen or sulphurous acid, instead of zinc or stannous chloride to reduce the iron, and the excess reducing agent removed by boiling, an accurate determination of the iron can be made.

PROCEDURE.—Grind the mineral to a fine powder. Weigh out two portions of about 0.5 gram each into small porcelain crucibles. Roast the ore at dull redness for ten minutes (Note 1), allow the crucibles to cool, and place them and their contents in casseroles containing 30 cc. of dilute hydrochloric acid (sp. gr. 1.12).

Proceed with the solution of the ore, and the treatment of the residue, if necessary, exactly as described for the bichromate process on page 56. When solution is complete, add 6 cc. of concentrated sulphuric acid to each casserole, and evaporate on the steam bath until the solution is nearly colorless (Note 2). Cover the casseroles and heat over the flame of the burner, holding the casserole in the hand and rotating it slowly to hasten evaporation and prevent spattering, until the heavy white fumes of sulphuric anhydride are freely evolved (Note 3). Cool the casseroles, add 100 cc. of water (measured), and boil gently until the ferric sulphate is dissolved; pour the warm solution through the reductor which has been previously washed; proceed as described under standardization, taking pains to use the same volume and strength of acid and the same volume of wash-water as there prescribed, and titrate with the permanganate solution in the reductor flask, using the ferrous sulphate solution if the end-point should be overstepped.

From the corrected volume of permanganate solution used, calculate the percentage of iron (Fe) in the limonite.

[Note 1: The preliminary roasting is usually necessary because, even though the sulphuric acid would subsequently char the carbonaceous matter, certain nitrogenous bodies are not thereby rendered insoluble in the acid, and would be oxidized by the permanganate.]

[Note 2: The temperature of the steam bath is not sufficient to volatilize sulphuric acid. Solutions may, therefore, be left to evaporate overnight without danger of evaporation to dryness.]

[Note 3: The hydrochloric acid, both free and combined, is displaced by the less volatile sulphuric acid at its boiling point. Ferric sulphate separates at this point, since there is no water to hold it in solution and care is required to prevent bumping. The ferric sulphate usually has a silky appearance and is easily distinguished from the flocculent silica which often remains undissolved.]

!Zimmermann-Reinhardt Procedure!

!Method (B)!

PROCEDURE.—Grind the mineral to a fine powder. Weigh out two portions of about 0.5 gram each into small porcelain crucibles. Proceed with the solution of the ore, treat the residue, if necessary, and reduce the iron by the addition of stannous chloride, followed by mercuric chloride, as described for the bichromate process on page 56. Dilute the solution to about 400 cc. with cold water, add 10 cc. of the manganous sulphate titrating solution (Note 1, page 68) and titrate with the standard potassium permanganate solution to a faint pink (Note 1).

From the standardization data already obtained calculate the percentage of iron (Fe) in the limonite.

[Note 1: It has already been noted that hydrochloric acid reacts slowly in cold solutions with potassium permanganate. It is, however, possible to obtain a satisfactory, although somewhat fugitive end-point in the presence of manganous sulphate and phosphoric acid. The explanation of the part played by these reagents is somewhat obscure as yet. It is possible that an intermediate manganic compound is formed which reacts rapidly with the ferrous compounds—thus in effect catalyzing the oxidizing process.

While an excess of hydrochloric acid is necessary for the successful reduction of the iron by stannous chloride, too large an amount should be avoided in order to lessen the chance of reduction of the permanganate by the acid during titration.]

DETERMINATION OF THE OXIDIZING POWER OF PYROLUSITE

INDIRECT OXIDATION

Pyrolusite, when pure, consists of manganese dioxide. Its value as an oxidizing agent, and for the production of chlorine, depends upon the percentage of MnO_2 in the sample. This percentage is determined by an indirect method, in which the manganese dioxide is reduced and dissolved by an excess of ferrous sulphate or oxalic acid in the presence of sulphuric acid, and the unused excess determined by titration with standard permanganate solution.

PROCEDURE.—Grind the mineral in an agate mortar until no grit whatever can be detected under the pestle (Note 1). Transfer it to a stoppered weighing-tube, and weigh out two portions of about 0.5 gram into beakers (400-500 cc.) Read Note 2, and then calculate in each case the weight of oxalic acid ($H_2C_2O_4.2H_2O$) required to react with the weights of pyrolusite taken. The reaction involved is

$$MnO_2 + H_2C_2O_4(2H_2O) + H_2SO_4 \rightarrow MnSO_4 + 2CO_2 + 4H_2O.$$

Weigh out about 0.2 gram in excess of this quantity of !pure! oxalic acid into the corresponding beakers, weighing the acid accurately and recording the weight in the notebook. Pour into each beaker 25 cc. of water and 50 cc. of dilute sulphuric acid (1:5), cover and warm the beaker and its contents gently until the evolution of carbon dioxide ceases (Note 3). If a residue remains which is sufficiently colored to obscure the end-reaction of the permanganate, it must be removed by filtration.

Finally, dilute the solution to 200-300 cc., heat the solution to a temperature just below boiling, add 15 cc. of a manganese sulphate solution and while hot, titrate for the excess of the oxalic acid with standard permanganate solution (Notes 4 and 5).

From the corrected volume of the solution required, calculate the amount of oxalic acid undecomposed by the pyrolusite; subtract this from the total quantity of acid used, and calculate the weight of manganese dioxide which would react with the balance of the acid, and from this the percentage in the sample.

[Note 1: The success of the analysis is largely dependent upon the fineness of the powdered mineral. If properly ground, solution should be complete in fifteen minutes or less.]

[Note 2: A moderate excess of oxalic acid above that required to react with the pyrolusite is necessary to promote solution; otherwise the residual quantity of oxalic acid would be so small that the last particles of the mineral would scarcely dissolve. It is also desirable that a sufficient excess of the acid should be present to react with a considerable volume of the permanganate solution during the titration, thus increasing the accuracy of the process. On the other hand, the excess of oxalic acid should not be so large as to react with more of the permanganate solution than is contained in a 50 cc. burette. If the pyrolusite under examination is known to be of high grade, say 80 per cent pure, or above the calculation of the oxalic acid needed may be based upon an assumption that the mineral is all MnO_2. If the quality of the mineral is unknown, it is better to weigh out three portions instead of two and to add to one of these the amount of oxalic prescribed, assuming complete purity of the mineral. Then run in the permanganate solution from a pipette or burette to determine roughly the amount required. If the volume exceeds the contents of a burette, the amount of oxalic acid added to the other two portions is reduced accordingly.]

[Note 3: Care should be taken that the sides of the beaker are not overheated, as oxalic acid would be decomposed by heat alone if crystallization should occur on the sides of the vessel. Strong sulphuric acid also decomposes the oxalic acid. The dilute acid should, therefore, be prepared before it is poured into the beaker.]

[Note 4: Ferrous ammonium sulphate, ferrous sulphate, or iron wire may be substituted for the oxalic acid. The reaction is then the following:

$$2\ FeSO_4 + MnO_2 + 2H_2SO_4 \longrightarrow Fe_2(SO_4)_3 + 2H_2O$$

The excess of ferrous iron may also be determined by titration with potassium bichromate, if desired. Care is required to prevent the oxidation of the iron by the air, if ferrous salts are employed.]

[Note 5: The oxidizing power of pyrolusite may be determined by other volumetric processes, one of which is outlined in the following reactions:

$MnO_2 + 4HCl \longrightarrow MnCl_2 + Cl_2 + 2H_2O$
$Cl_2 + 2KI \longrightarrow I_2 + 2KCl$
$I_2 + 2Na_2S_2O_3 \longrightarrow Na_2S_4O_6 + 2NaI.$

The chlorine generated by the pyrolusite is passed into a solution of potassium iodide. The liberated iodine is then determined by titration with sodium thiosulphate, as described on page 78. This is a direct process, although it involves three steps.]

IODIMETRY

The titration of iodine against sodium thiosulphate, with starch as an indicator, may perhaps be regarded as the most accurate of volumetric processes. The thiosulphate solution may be used in both acid and neutral solutions to measure free iodine and the latter may, in turn, serve as a measure of any substance capable of liberating iodine from potassium iodide under suitable conditions for titration, as, for example, in the process outlined in Note 5 on page 74.

The fundamental reaction upon which iodometric processes are based is the following:

$I_2 + 2 Na_2S_2O_3 \longrightarrow 2 NaI + Na_2S_4O_6$.

This reaction between iodine and sodium thiosulphate, resulting in the formation of the compound $Na_2S_4O_6$, called sodium tetrathionate, is quantitatively exact, and differs in that respect from the action of chlorine or bromine, which oxidize the thiosulphate, but not quantitatively.

NORMAL SOLUTIONS OF IODINE AND SODIUM THIOSULPHATE

If the formulas of sodium thiosulphate and sodium tetrathionate are written in a manner to show the atoms of oxygen associated with sulphur atoms in each, thus, $2(Na_2).S_2O_2$ and $Na_2O.S_4O_5$, it is plain that in the tetrathionate there are five atoms of oxygen associated with sulphur, instead of the four in the two molecules of the thiosulphate taken together. Although, therefore, the iodine contains no oxygen, the two atoms of iodine have, in effect, brought about the addition of one oxygen atoms to the sulphur atoms. That is the same thing as saying that 253.84 grams of iodine (I_2) are equivalent to 16 grams of oxygen; hence, since 8 grams of oxygen is the basis of normal solutions, 253.84/2 or 126.97 grams of iodine should be contained in one liter of normal iodine solution. By a similar course of reasoning the conclusion is reached that the normal solution of sodium thiosulphate should contain, per liter, its molecular weight in grams. As the thiosulphate in crystalline form has the formula $Na_2S_2O_3.5H_2O$, this weight is 248.12 grams. Tenth-normal or hundredth-normal solutions are generally used.

PREPARATION OF STANDARD SOLUTIONS

!Approximate Strength, 0.1 N!

PROCEDURE.—Weigh out on the rough balances 13 grams of commercial iodine. Place it in a mortar with 18 grams of potassium iodide and triturate with small portions of water until all is dissolved. Dilute the solution to 1000 cc. and transfer to a liter bottle and mix thoroughly (Note 1).[1]

[Footnote 1: It will be found more economical to have a considerable quantity of the solution prepared by a laboratory attendant, and to have all unused solutions returned to the common stock.]

Weigh out 25 grams of sodium thiosulphate, dissolve it in water which has been previously boiled and cooled, and dilute to 1000 cc., also with boiled water. Transfer the solution to a liter bottle and mix thoroughly (Note 2).

[Note 1: Iodine solutions react with water to form hydriodic acid under the influence of the sunlight, and even at low room temperatures the iodine tends to volatilize from solution. They should, therefore, be protected from light and heat. Iodine solutions are not stable for long periods under the best of conditions. They cannot be used in burettes with rubber tips, since they attack the rubber.]

[Note 2: Sodium thiosulphate ($Na_2S_2O_3.5H_2O$) is rarely wholly pure as sold commercially, but may be purified by recrystallization. The carbon dioxide absorbed from the air by distilled water decomposes the salt, with the separation of sulphur. Boiled water which has been cooled out of contact with the air should be used in preparing solutions.]

INDICATOR SOLUTION

The starch solution for use as an indicator must be freshly prepared. A soluble starch is obtainable which serves well, and a solution of 0.5 gram of this starch in 25 cc. of boiling water is sufficient. The solution should be filtered while hot and is ready for use when cold.

If soluble starch is not at hand, potato starch may be used. Mix about 1 gram with 5 cc. of cold water to a smooth paste, pour 150 cc. of !boiling! water over it, warm for a moment on the hot plate, and put it aside to settle. Decant the supernatant liquid through a filter and use the clear filtrate; 5 cc. of this solution are needed for a titration.

The solution of potato starch is less stable than the soluble starch. The solid particles of the starch, if not removed by filtration, become so colored by the iodine that they are not readily decolorized by the thiosulphate (Note 1).

[Note 1: The blue color which results when free iodine and starch are brought together is probably not due to the formation of a true chemical compound. It is regarded as a "solid solution" of iodine in starch. Although it is unstable, and easily destroyed by heat, it serves as an indicator for the presence of free iodine of remarkable sensitiveness, and makes the iodometric processes the most satisfactory of any in the field of volumetric analysis.]

COMPARISON OF IODINE AND THIOSULPHATE SOLUTIONS

PROCEDURE.—Place the solutions in burettes (the iodine in a glass-stoppered burette), observing the usual precautions. Run out 40 cc. of the thiosulphate solution into a beaker, dilute with 150 cc. of water, add 1 cc. to 2 cc. of the soluble starch solution, and titrate with the iodine to the appearance of the blue of the iodo-starch. Repeat until the ratio of the two solutions is established, remembering all necessary corrections for burettes and for temperature changes.

STANDARDIZATION OF SOLUTIONS

Commercial iodine is usually not sufficiently pure to permit of its use as a standard for thiosulphate solutions or the direct preparation of a standard solution of iodine. It is likely to contain, beside moisture, some iodine chloride, if chlorine was used to liberate the iodine when it was prepared. It may be purified by sublimation after mixing it with a little potassium iodide, which reacts with the iodine chloride, forming potassium chloride and setting free the iodine. The sublimed iodine is then dried by placing it in a closed container over concentrated sulphuric acid. It may then be weighed in a stoppered weighing-tube and dissolved in a solution of potassium iodide in a stoppered flask to prevent loss of iodine by volatilization. About 18 grams of the iodide and twelve grams of iodine per liter are required for an approximately tenth-normal solution.

An iodine solution made from commercial iodine may also be standardized against arsenious oxide (As_4O_6). This substance also usually requires purification by sublimation before use.

The substances usually employed for the standardization of a thiosulphate solution are potassium bromate and metallic copper. The former is obtainable in pure condition or may be easily purified by re-crystallization. Copper wire of high grade is sufficiently pure to serve as a standard. Both potassium bromate and cupric salts in solution will liberate iodine from an iodide, which is then titrated with the thiosulphate solution.

The reactions involved are the following:

(a) $KBrO_3 + 6KI + 3H_2SO_4 \longrightarrow KBr + 3I_2 + 3K_2SO_4 + 3H_2O$,

(b) $3Cu + 8HNO_3 \longrightarrow 3Cu(NO_3)_2 + 2NO + 4H_2O$, $2Cu(NO_3)_2 + 4KI \longrightarrow 2CuI + 4KNO_3 + I_2$.

Two methods for the direct standardization of the sodium thiosulphate solution are here described, and one for the direct standardization of the iodine solution.

!Method A!

PROCEDURE.—Weigh out into 500 cc. beakers two portions of about 0.150-0.175 gram of potassium bromate. Dissolve each of these in 50 cc. of water, and add 10 cc. of a potassium iodide solution containing 3 grams of the salt in that volume (Note 1). Add to the mixture 10 cc. of dilute sulphuric acid (1 volume of sulphuric acid with 5 volumes of water), allow the solution to stand for three minutes, and dilute to 150 cc. (Note 2). Run in thiosulphate solution from a burette until the color of the liberated iodine is nearly destroyed, and then add 1 cc. or 2 cc. of starch solution, titrate to the disappearance of the iodo-starch blue, and finally add iodine solution until the color is just restored. Make a blank test for the amount of thiosulphate solution required to react with the iodine liberated by the iodate which is generally present in the potassium iodide solution, and deduct this from the total volume used in the titration.

From the data obtained, calculate the relation of the thiosulphate solution to a normal solution, and subsequently calculate the similar value for the iodine solution.

[Note 1:—Potassium iodide usually contains small amounts of potassium iodate as impurity which, when the iodide is brought into an acid solution, liberates iodine, just as does the potassium bromate used as a standard. It is necessary to determine the amount of thiosulphate which reacts with the iodine thus liberated by making a "blank test" with the iodide and acid alone. As the iodate is not always uniformly distributed throughout the iodide, it is better to make up a sufficient volume of a solution of the iodide for the purposes of the work in hand, and to make the blank test by using the same volume of the iodide solution as is added in the standardizing process. The iodide solution should contain about 3 grams of the salt in 10 cc.]

[Note 2: The color of the iodo-starch is somewhat less satisfactory in concentrated solutions of the alkali salts, notably the iodides. The dilution prescribed obviates this difficulty.]

!Method B!

PROCEDURE.—Weigh out two portions of 0.25-0.27 gram of clean copper wire into 250 cc. Erlenmeyer flasks (Note 1). Add to each 5 cc. of concentrated nitric acid (sp. gr. 1.42) and 25 cc. of water, cover, and warm until solution is complete. Add 5 cc. of bromine water and boil until the excess of bromine is expelled. Cool, and add strong ammonia (sp. gr. 0.90) drop by drop until a deep blue color indicates the presence of an excess. Boil the solution until the deep blue is replaced by a light bluish green, or a brown stain appears on the sides of the flask (Note 2). Add 10 cc. of strong acetic acid (sp. gr. 1.04), cool under the water tap, and add a solution of potassium iodide (Note 3) containing about 3 grams of the salt, and titrate with thiosulphate solution until the color of the liberated iodine is nearly destroyed. Then add 1-2 cc. of freshly prepared starch solution, and add thiosulphate solution, drop by drop, until the blue color is discharged.

From the data obtained, including the "blank test" of the iodide, calculate the relation of the thiosulphate solution to the normal.

[Note 1: While copper wire of commerce is not absolutely pure, the requirements for its use as a conductor of electricity are such that the impurities constitute only a few hundredths of one per cent and are negligible for analytical purposes.]

[Note 2: Ammonia neutralizes the free nitric acid. It should be added in slight excess only, since the excess must be removed by boiling, which is tedious. If too much ammonia is present when acetic acid is added, the resulting ammonium acetate is hydrolyzed, and the ammonium hydroxide reacts with the iodine set free.]

[Note 3: A considerable excess of potassium iodide is necessary for the prompt liberation of iodine. While a large excess will do no harm, the cost of this reagent is so great that waste should be avoided.]

!Method C!

PROCEDURE.—Weigh out into 500 cc. beakers two portions of 0.175-0.200 gram each of pure arsenious oxide. Dissolve each of these in 10 cc. of sodium hydroxide solution, with stirring. Dilute the solutions to 150 cc. and add dilute hydrochloric acid until the solutions contain a few drops in excess, and finally add to each a concentrated solution of 5 grams of pure sodium bicarbonate (Na-

HCO_3) in water. Cover the beakers before adding the bicarbonate, to avoid loss. Add the starch solution and titrate with the iodine to the appearance of the blue of the iodo-starch, taking care not to pass the end-point by more than a few drops (Note 1).

From the corrected volume of the iodine solution used to oxidize the arsenious oxide, calculate its relation to the normal. From the ratio between the solutions, calculate the similar value for the thiosulphate solution.

[Note 1: Arsenious oxide dissolves more readily in caustic alkali than in a bicarbonate solution, but the presence of caustic alkali during the titration is not admissible. It is therefore destroyed by the addition of acid, and the solution is then made neutral with the solution of bicarbonate, part of which reacts with the acid, the excess remaining in solution.

The reaction during titration is the following:

$$Na_3AsO_3 + I_2 + 2NaHCO_3 \longrightarrow Na_3AsO_4 + 2NaI + 2CO_2 + H_2O$$

As the reaction between sodium thiosulphate and iodine is not always free from secondary reactions in the presence of even the weakly alkaline bicarbonate, it is best to avoid the addition of any considerable excess of iodine. Should the end-point be passed by a few drops, the thiosulphate may be used to correct it.]

DETERMINATION OF COPPER IN ORES

Copper ores vary widely in composition from the nearly pure copper minerals, such as malachite and copper sulphide, to very low grade materials which contain such impurities as silica, lead, iron, silver, sulphur, arsenic, and antimony. In nearly all varieties there will be found a siliceous residue insoluble in acids. The method here given, which is a modification of that described by A.H. Low (!J. Am. Chem. Soc.! (1902), 24, 1082), provides for the extraction of the copper from commonly occurring ores, and for the presence of their common impurities. For practice analyses it is advisable to select an ore of a fair degree of purity.

PROCEDURE.— Weigh out two portions of about 0.5 gram each of the ore (which should be ground until no grit is detected) into 250 cc. Erlenmeyer flasks or small beakers. Add 10 cc. of concentrated nitric acid (sp. gr. 1.42) and heat very gently until the ore is decomposed and the acid evaporated nearly to dryness (Note 1). Add 5 cc. of concentrated hydrochloric acid (sp. gr. 1.2) and warm gently. Then add about 7 cc. of concentrated sulphuric acid (sp. gr. 1.84) and evaporate over a free flame until the sulphuric acid fumes freely (Note 2). It has then displaced nitric and hydrochloric acid from their compounds.

Cool the flask or beaker, add 25 cc. of water, heat the solution to boiling, and boil for two minutes. Filter to remove insoluble sulphates, silica and any silver that may have been precipitated as silver chloride, and receive the filtrate in a small beaker, washing the precipitate and filter paper with warm water until the filtrate and washings amount to 75 cc. Bend a strip of aluminium foil (5 cm. x 12 cm.) into triangular form and place it on edge in the beaker. Cover the beaker and boil the solution (being careful to avoid loss of liquid by spattering) for ten minutes, but do not evaporate to small volume.

Wash the cover glass and sides of the beaker. The copper should now be in the form of a precipitate at the bottom of the beaker or adhering loosely to the aluminium sheet. Remove the sheet, wash it carefully with hydrogen sulphide water and place it in a small beaker. Decant the solution through a filter, wash the precipitated copper twice by decantation with hydrogen sulphide water, and finally transfer the copper to the filter paper, where it is again washed thoroughly, being careful at all times to keep the precipitated copper covered with the wash water. Remove and discard the filtrate and place an Erlenmeyer flask under the funnel. Pour 15 cc. of dilute nitric acid (sp. gr. 1.20) over the aluminium foil in the beaker, thus dissolving any adhering copper. Wash the foil with hot water and remove it. Warm this nitric acid solution and pour it slowly through the filter paper, thereby dissolving the copper on the paper, receiving the acid solution in the Erlenmeyer flask. Before washing the paper, pour 5 cc. of saturated bromine water (Note 3) through it and finally wash the paper carefully with hot water and transfer any particles of copper which may be left on it to the Erlenmeyer flask. Boil to expel the bromine. Add concentrated ammonia drop by drop until the appearance of a deep blue coloration indicates an excess. Boil until the deep blue is displaced by a light bluish green coloration, or until brown stains form on the sides of the flask. Add 10 cc. of strong acetic acid (Note 4) and cool under the water tap. Add a solution containing about 3 grams of potassium iodide, as in the standardization, and titrate with thiosulphate solution until the yellow of the liberated iodine is nearly discharged. Add 1-2 cc. of freshly prepared starch solution and titrate to the disappearance of the blue color.

From the data obtained, calculate the percentage of copper (Cu) in the ore.

[Note 1: Nitric acid, because of its oxidizing power, is used as a solvent for the sulphide ores. As a strong acid it will also dissolve the copper from carbonate ores. The hydrochloric acid is added to dissolve oxides of iron and to precipitate silver and lead. The sulphuric acid displaces the other acids, leaving a solution containing sulphates only. It also, by its dehydrating action, renders silica from silicates insoluble.]

[Note 2: Unless proper precautions are taken to insure the correct concentrations of acid the copper will not precipitate quantitatively on the aluminium foil; hence care must be taken to follow directions carefully at this point. Lead and silver have been almost completely removed as sulphate and chloride respectively, or they too would be precipitated on the aluminium. Bismuth, though precipitated on aluminium, has no effect on the analysis. Arsenic and antimony precipitate on aluminium and would interfere with the titration if allowed to remain in the lower state of oxidation.]

[Note 3: Bromine is added to oxidize arsenious and antimonious compounds from the original sample, and to oxidize nitrous acid formed by the action of nitric acid on copper and copper sulphide.]

[Note 4: This reaction can be carried out in the presence of sulphuric and hydrochloric acids as well as acetic acid, but in the presence of these strong acids arsenic and antimonic acids may react with the hydriodic acid produced with the liberation of free iodine, thereby reversing the process and introducing an error.]

DETERMINATION OF ANTIMONY IN STIBNITE

Stibnite is native antimony sulphide. Nearly pure samples of this mineral are easily obtainable and should be used for practice, since many impurities, notably iron, seriously interfere with the accurate determination of the antimony by iodometric methods. It is, moreover, essential that the directions with respect to amounts of reagents employed and concentration of solutions should be followed closely.

PROCEDURE.—Grind the mineral with great care, and weigh out two portions of 0.35-0.40 gram into small, dry beakers (100 cc.). Cover the beakers and pour over the stibnite 5 cc. of concentrated hydrochloric acid (sp. gr. 1.20) and warm gently on the water bath (Note 1). When the residue is white, add to each beaker 2 grams of powdered tartaric acid (Note 2). Warm the solution on the water bath for ten minutes longer, dilute the solution very cautiously by adding water in portions of 5 cc., stopping if the solution turns red. It is possible that no coloration will appear, in which case cautiously continue the dilution to 125 cc. If a red precipitate or coloration does appear, warm the solution until it is colorless, and again dilute cautiously to a total volume of 125 cc. and boil for a minute (Note 3).

If a white precipitate of the oxychloride separates during dilution (which should not occur if the directions are followed), it is best to discard the determination and to start anew.

Carefully neutralize most of the acid with ammonium hydroxide solution (sp. gr. 0.96), but leave it distinctly acid (Note 4). Dissolve 3 grams of sodium bicarbonate in 200 cc. of water in a 500 cc. beaker, and pour the cold solution of the antimony chloride into this, avoiding loss by effervescence. Make sure that the solution contains an excess of the bicarbonate, and then add 1 cc. or 2 cc. of starch solution and titrate with iodine solution to the appearance of the blue, avoiding excess (Notes 5 and 6).

From the corrected volume of the iodine solution required to oxidize the antimony, calculate the percentage of antimony (Sb) in the stibnite.

[Note 1: Antimony chloride is volatile with steam from its concentrated solutions; hence these solutions must not be boiled until they have been diluted.]

[Note 2: Antimony salts, such as the chloride, are readily hydrolyzed, and compounds such as SbOCl are formed which are often relatively insoluble; but in the presence of tartaric acid compounds with complex ions are formed, and these are soluble. An excess of hydrochloric acid also prevents precipitation of the oxychloride because the H^{+} ions from the acid lessen the dissociation of the water and thus prevent any considerable hydrolysis.]

[Note 3: The action of hydrochloric acid upon the sulphide sets free sulphureted hydrogen, a part of which is held in solution by the acid. This is usually expelled by the heating upon the water bath; but if it is not wholly driven out, a point is reached during dilution at which the antimony sulphide, being no longer held in solution by the acid, separates. If the dilution is immediately stopped and the solution warmed, this sulphide is again brought into solution and at the same time more of the sulphureted hydrogen is expelled. This procedure must be continued until the sulphureted hydrogen is all removed, since it reacts with iodine. If no precipitation of the sulphide occurs, it is an indication that the sulphureted hydrogen was all expelled on solution of the stibnite.]

[Note 4: Ammonium hydroxide is added to neutralize most of the acid, thus lessening the amount of sodium bicarbonate to be added. The ammonia should not neutralize all of the acid.]

[Note 5: The reaction which takes place during titration may be expressed thus:

$Na_3SbO_3 + 2NaHCO_3 + I_2 \longrightarrow Na_3SbO_4 + 2NaI + H_2O + 2CO_2$.]

[Note 6: If the end-point is not permanent, that is, if the blue of the iodo-starch is discharged after standing a few moments, the cause may be an insufficient quantity of sodium bicarbonate, leaving the solution slightly acid, or a very slight precipitation of an antimony compound which is slowly acted upon by the iodine when the latter is momentarily present in excess. In either case it is better to discard the analysis and to repeat the process, using greater care in the amounts of reagents employed.]

CHLORIMETRY

The processes included under the term !chlorimetry! comprise those employed to determine chlorine, hypochlorites, bromine, and hypobromites. The reagent employed is sodium arsenite in the presence of sodium bicarbonate. The reaction in the case of the hypochlorites is

$NaClO + Na_3AsO_3 \longrightarrow Na_3AsO_4 + NaCl$.

The sodium arsenite may be prepared from pure arsenious oxide, as described below, and is stable for considerable periods; but commercial oxide requires resublimation to remove arsenic sulphide, which may be present in small quantity. To prepare the solution, dissolve about 5 grams of the powdered oxide, accurately weighed, in 10 cc. of a concentrated sodium hydroxide solution, dilute the solution to 300 cc., and make it faintly acid with dilute hydrochloric acid. Add 30 grams of sodium bicarbonate dissolved in a little water, and dilute the solution to exactly 1000 cc. in a measuring flask. Transfer the solution to a dry liter bottle and mix thoroughly.

It is possible to dissolve the arsenious oxide directly in a solution of sodium bicarbonate, with gentle warming, but solution in sodium hydroxide takes place much more rapidly, and the excess of the hydroxide is readily neutralized by hydrochloric acid, with subsequent addition of the bicarbonate to maintain neutrality during the titration.

The indicator required for this process is made by dipping strips of filter paper in a starch solution prepared as described on page 76, to which 1 gram of potassium iodide has been added. These strips are allowed to drain and spread upon a watch-glass until dry. When touched by a drop of the solution the paper turns blue until the hypochlorite has all been reduced and an excess of the arsenite has been added.

DETERMINATION OF THE AVAILABLE CHLORINE IN BLEACHING POWDER

Bleaching powder consists mainly of a calcium compound which is a derivative of both hydrochloric and hypochlorous acids. Its formula is CaClOCl. Its use as a bleaching or disinfecting agent, or as a source of chlorine, depends upon the amount of hypochlorous acid which it yields when treated with a stronger acid. It is customary to express the value of bleaching powder in terms of "available chlorine," by which is meant the chlorine present as hypochlorite, but not the chlorine present as chloride.

PROCEDURE.—Weigh out from a stoppered test tube into a porcelain mortar about 3.5 grams of bleaching powder (Note 1). Triturate the powder in the mortar with successive portions of water until it is well ground and wash the contents into a 500 cc. measuring flask (Note 2). Fill the flask to the mark with water and shake thoroughly. Measure off 25 cc. of this semi-solution in a measuring flask, or pipette, observing the precaution that the liquid removed shall contain approximately its proportion of suspended matter.

Empty the flask or pipette into a beaker and wash it out. Run in the arsenite solution from a burette until no further reaction takes place on the starch-iodide paper when touched by a drop of the solution of bleaching powder. Repeat the titration, using a second 25 cc. portion.

From the volume of solution required to react with the bleaching powder, calculate the percentage of available chlorine in the latter, assuming the titration reaction to be that between chlorine and arsenious oxide:

$As_4O_6 + 4Cl_2 + 4H_2O \longrightarrow 2As_2O_5 + 8HCl$

Note that only one twentieth of the original weight of bleaching powder enters into the reaction.

[Note 1: The powder must be triturated until it is fine, otherwise the lumps will inclose calcium hypochlorite, which will fail to react with the arsenious acid. The clear supernatant liquid gives percentages which are below, and the sediment percentages which are above, the average. The liquid measured off should, therefore, carry with it its proper proportion of the sediment, so far as that can be brought about by shaking the solution just before removal of the aliquot part for titration.]

[Note 2: Bleaching powder is easily acted upon by the carbonic acid in the air, which liberates the weak hypochlorous acid. This, of course, results in a loss of available chlorine. The original material for analysis should be kept in a closed container and protected form the air as far as possible. It is difficult to obtain analytical samples which are accurately representative of a large quantity of the bleaching powder. The procedure, as outlined, will yield results which are sufficiently exact for technical purposes.]

III. PRECIPITATION METHODS

DETERMINATION OF SILVER BY THE THIOCYANATE PROCESS

The addition of a solution of potassium or ammonium thiocyanate to one of silver in nitric acid causes a deposition of silver thiocyanate as a white, curdy precipitate. If ferric nitrate is also present, the slightest excess of the thiocyanate over that required to combine with the silver is indicated by the deep red which is characteristic of the thiocyanate test for iron.

The reactions involved are:

$AgNO_3 + KSCN \rightarrow AgSCN + KNO_3$, $3KSCN + Fe(NO_3)_3 \rightarrow Fe(SCN)_3 + 3KNO_3$.

The ferric thiocyanate differs from the great majority of salts in that it is but very little dissociated in aqueous solutions, and the characteristic color appears to be occasioned by the formation of the un-ionized ferric salt.

The normal solution of potassium thiocyanate should contain an amount of the salt per liter of solution which would yield sufficient $(CNS)^{-}$ to combine with one gram of hydrogen to form HCNS, i.e., a gram-molecular weight of the salt or 97.17 grams. If the ammonium thiocyanate is used, the amount is 76.08 grams. To prepare the solution for this determination, which should be approximately 0.05 N, dissolve about 5 grams of potassium thiocyanate, or 4 grams of ammonium thiocyanate, in a small amount of water; dilute this solution to 1000 cc. in a liter bottle and mix as usual.

Prepare 20 cc. of a saturated solution of ferric alum and add 5 cc. of dilute nitric acid (sp. gr. 1.20). About 5 cc. of this solution should be used as an indicator.

STANDARDIZATION

PROCEDURE.—Crush a small quantity of silver nitrate crystals in a mortar (Note 1). Transfer them to a watch-glass and dry them for an hour at 110°C., protecting them from dust or other organic matter (Note 2). Weigh out two portions of about 0.5 gram each and dissolve them in 50 cc. of water. Add 10 cc. of dilute nitric acid which has been recently boiled to expel the lower oxides of nitrogen, if any, and then add 5 cc. of the indicator solution. Run in the thiocyanate solution from a burette, with constant stirring, allowing the precipitate to settle occasionally to obtain an exact recognition of the end-point, until a faint red tinge can be detected in the solution.

From the data obtained, calculate the relation of the thiocyanate solution to the normal.

[Note 1: The thiocyanate cannot be accurately weighed; its solutions must, therefore, be standardized against silver nitrate (or pure silver), either in the form of a standard solution or in small, weighed portions.]

[Note 2: The crystals of silver nitrate sometimes inclose water which is expelled on drying. If the nitrate has come into contact with organic bodies it suffers a reduction and blackens during the heating.

It is plain that a standard solution of silver nitrate (made by weighing out the crystals) is convenient or necessary if many titrations of this nature are to be made. In the absence of such a solution the liability of passing the end-point is lessened by setting aside a small fraction of the silver solution, to be added near the close of the titration.]

DETERMINATION OF SILVER IN COIN

PROCEDURE.— Weigh out two portions of the coin of about 0.5 gram each. Dissolve them in 15 cc. of dilute nitric acid (sp. gr. 1.2) and boil until all the nitrous compounds are expelled (Note 1). Cool the solution, dilute to 50 cc., and add 5 cc. of the indicator solution, and titrate with the thiocyanate to the appearance of the faint red coloration (Note 2).

From the corrected volume of the thiocyanate solution required, calculate the percentage of silver in the coin.

[Note 1: The reaction with silver may be carried out in nitric acid solutions and in the presence of copper, if the latter does not exceed 70 per cent. Above that percentage it is necessary to add silver in known quantity to the solution. The liquid must be cold at the time of titration and entirely free from nitrous compounds, as these sometimes cause a reddening of the indicator solution. All utensils, distilled water, the nitric acid and the beakers must be free from chlorides, as the presence of these will cause precipitation of silver chloride, thereby introducing an error.]

[Note 2: The solution containing the silver precipitate, as well as those from the standardization, should be placed in the receptacle for "silver residues" as a matter of economy.]

PART III

GRAVIMETRIC ANALYSIS

GENERAL DIRECTIONS

Gravimetric analyses involve the following principal steps: first, the weighing of the sample; second, the solution of the sample; third, the separation of some substance from solution containing, or bearing a definite relation to, the constituent to be measured, under conditions which render this separation as complete as possible; and finally, the segregation of that substance, commonly by filtration, and the determination of its weight, or that of some stable product formed from it on ignition. For example, the gravimetric determination of aluminium is accomplished by solution of the sample, by precipitation in the form of hydroxide, collection of the hydroxide upon a filter, complete removal by washing of all foreign soluble matter, and the burning of the filter and ignition of the precipitate to aluminium oxide, in which condition it is weighed.

Among the operations which are common to nearly all gravimetric analyses are precipitation, washing of precipitates, ignition of precipitates, and the use of desiccators. In order to avoid burdensome repetitions in the descriptions of the various gravimetric procedures which follow, certain general instructions are introduced at this point. These instructions must, therefore, be considered to be as much a part of all subsequent procedures as the description of apparatus, reagents, or manipulations.

The analytical balance, the fundamentally important instrument in gravimetric analysis, has already been described on pages 11 to 15.

PRECIPITATION

For successful quantitative precipitations those substances are selected which are least soluble under conditions which can be easily established, and which separate from solution in such a state that they can be filtered readily and washed free from admixed material.

In general, the substances selected are the same as those already familiar to the student of Qualitative Analysis.

When possible, substances are selected which separate in crystalline form, since such substances are less likely to clog the pores of filter paper and can be most quickly washed. In order to increase the size of the crystals, which further promotes filtration and washing, it is often desirable to allow a precipitate to remain for some time in contact with the solution from which it has separated. The solution is often kept warm during this period of "digestion." The small crystals gradually disappear and the larger crystals increase in size, probably as the result of the force known as surface tension, which tends to reduce the surface of a given mass of material to a minimum, combined with a very slightly greater solubility of small crystals as compared with the larger ones.

Amorphous substances, such as ferric hydroxide, aluminium hydroxide, or silicic acid, separate in a gelatinous form and are relatively difficult to filter and wash. Substances of this class also exhibit a tendency to form, with pure water, what are known as colloidal solutions. To prevent this as far as possible, they are washed with solutions of volatile salts, as will be described in some of the following procedures.

In all precipitations the reagent should be added slowly, with constant stirring, and should be hot when circumstances permit. The slow addition is less likely to occasion contamination of the precipitate by the inclosure of other substances which may be in the solution, or of the reagent itself.

FUNNELS AND FILTERS

Filtration in analytical processes is most commonly effected through paper filters. In special cases these may be advantageously replaced by an asbestos filter in a perforated porcelain or platinum crucible, commonly known, from its originator, as a "Gooch filter." The operation and use of a filter of this type is described on page 103. Porous crucibles of a material known as alundum may also be employed to advantage in special cases.

The glass funnels selected for use with paper filters should have an angle as near 60° as possible, and a narrow stem about six inches in length. The filters employed should be washed filters, i.e., those which have been treated with hydrochloric and hydrofluoric acids, and which on incineration leave a very small and definitely known weight of ash, generally about .00003 gram. Such filters are readily obtainable on the market.

The filter should be carefully folded to fit the funnel according to either of the two well-established methods described in the Appendix. It should always be placed so that the upper edge of the paper is about one fourth inch below the top of the funnel. Under no circumstances should the filter extend above the edge of the funnel, as it is then utterly impossible to effect complete washing.

To test the efficiency of the filter, fill it with distilled water. This water should soon fill the stem completely, forming a continuous column of liquid which, by its hydrostatic pressure, produces a gentle suction, thus materially promoting the rapidity of filtration. Unless the filter allows free passage of water under these conditions, it is likely to give much trouble when a precipitate is placed upon it.

The use of a suction pump to promote filtration is rarely altogether advantageous in quantitative analysis, if paper filters are employed. The tendency of the filter to break, unless the point of the filter paper is supported by a perforated porcelain cone or a small "hardened filter" of parchment, and the tendency of the precipitates to pass through the pores of the filter, more than compensate for the possible gain in time. On the other hand, filtration by suction may be useful in the case of precipitates which do not require ignition before weighing, or in the case of precipitates which are to be discarded without weighing. This is best accomplished with the aid of the special apparatus called a Gooch filter referred to above.

FILTRATION AND WASHING OF PRECIPITATES

Solutions should be filtered while hot, as far as possible, since the passage of a liquid through the pores of a filter is retarded by friction, and this, for water at 100°C., is less than one sixth of the resistance at 0°C.

When the filtrate is received in a beaker, the stem of the funnel should touch the side of the receiving vessel to avoid loss by spattering. Neglect of this precaution is a frequent source of error.

The vessels which contain the initial filtrate should !always! be replaced by clean ones, properly labeled, before the washing of a precipitate begins. In many instances a finely divided precipitate which shows no tendency to pass through the filter at first, while the solution is relatively dense, appears at once in the washings. Under such conditions the advantages accruing from the removal of the first filtrate are obvious, both as regards the diminished volume requiring refiltration, and also the smaller number of washings subsequently required.

Much time may often be saved by washing precipitates by decantation, i.e., by pouring over them, while still in the original vessel, considerable volumes of wash-water and allowing them to settle. The supernatant, clear wash-water is then decanted through the filter, so far as practicable without disturbing the precipitate, and a new portion of wash-water is added. This procedure can be employed to special advantage with gelatinous precipitates, which fill up the pores of the filter paper. As the medium from which the precipitate is to settle becomes less dense it subsides less readily, and it ultimately becomes necessary to transfer it to the filter and complete the washing there.

A precipitate should never completely fill a filter. The wash-water should be applied at the top of the filter, above the precipitate. It may be shown mathematically that the washing is most !rapidly! accomplished by filling the filter well to the top with wash-water each time, and allowing it to drain completely after each addition; but that when a precipitate is to be washed with the !least possible volume! of liquid the latter should be applied in repeated !small! quantities.

Gelatinous precipitates should not be allowed to dry before complete removal of foreign matter is effected. They are likely to shrink and crack, and subsequent additions of wash-water pass through these channels only.

All filtrates and wash-waters without exception must be properly tested. !This lies at the foundation of accurate work!, and the student should clearly understand that it is only by the invariable application of this rule that assurance of ultimate reliability can be secured. Every original filtrate must be tested to prove complete precipitation of the compound to be separated, and the wash-waters must also be tested to assure complete removal of foreign material. In testing the latter, the amount first taken should be but a few drops if the filtrate contains material which is to be subsequently determined. When, however, the washing of the filter and precipitate is nearly completed the amount should be increased, and for the final test not less than 3 cc. should be used.

It is impossible to trust to one's judgment with regard to the washing of precipitates; the washings from !each precipitate! of a series simultaneously treated must be tested, since the rate of washing will often differ materially under apparently similar conditions, !No exception can ever be made to this rule!.

The habit of placing a clean common filter paper under the receiving beaker during filtration is one to be commended. On this paper a record of the number of washings can very well be made as the portions of wash-water are added.

It is an excellent practice, when possible, to retain filtrates and precipitates until the completion of an analysis, in order that, in case of question, they may be examined to discover sources of error.

For the complete removal of precipitates from containing vessels, it is often necessary to rub the sides of these vessels to loosen the adhering particles. This can best be done by slipping over the end of a stirring rod a soft rubber device sometimes called a "policeman."

DESICCATORS

Desiccators should be filled with fused, anhydrous calcium chloride, over which is placed a clay triangle, or an iron triangle covered with silica tubes, to support the crucible or other utensils. The cover of the desiccator should be made air-tight by the use of a thin coating of vaseline.

Pumice moistened with concentrated sulphuric acid may be used in place of the calcium chloride, and is essential in special cases; but for most purposes the calcium chloride, if renewed occasionally and not allowed to cake together, is practically efficient and does not slop about when the desiccator is moved.

Desiccators should never remain uncovered for any length of time. The dehydrating agents rapidly lose their efficiency on exposure to the air.

CRUCIBLES

It is often necessary in quantitative analysis to employ fluxes to bring into solution substances which are not dissolved by acids. The fluxes in most common use are sodium carbonate and sodium or potassium acid sulphate. In gravimetric analysis it is usually necessary to ignite the separated substance after filtration and washing, in order to remove moisture, or to convert it through physical or chemical changes into some definite and stable form for weighing. Crucibles to be used in fusion processes must be made of materials which will withstand the action of the fluxes employed, and crucibles to be used for ignitions must be made of material which will not undergo any permanent change during the ignition, since the initial weight of the crucible must be deducted from the final weight of the crucible and product to obtain the weight of the ignited substance. The three materials which satisfy these conditions, in general, are platinum, porcelain, and silica.

Platinum crucibles have the advantage that they can be employed at high temperatures, but, on the other hand, these crucibles can never be used when there is a possibility of the reduction to the metallic state of metals like lead, copper, silver, or gold, which would alloy with and ruin the crucible. When platinum crucibles are used with compounds of arsenic or phosphorus, special precautions are necessary to prevent damage. This statement applies to both fusions and ignitions.

Fusions with sodium carbonate can be made only in platinum, since porcelain or silica crucibles are attacked by this reagent. Acid sulphate fusions, which require comparatively low temperatures, can sometimes be made in platinum, although platinum is slightly attacked by the flux. Porcelain or silica crucibles may be used with acid fluxes.

Silica crucibles are less likely to crack on heating than porcelain crucibles on account of their smaller coefficient of expansion. Ignition of substances not requiring too high a temperature may be made in porcelain or silica crucibles.

Iron, nickel or silver crucibles are used in special cases.

In general, platinum crucibles should be used whenever such use is practicable, and this is the custom in private, research or commercial laboratories. Platinum has, however, become so valuable that it is liable to theft unless constantly under the protection of the user. As constant protection is often difficult in instructional laboratories, it is advisable, in order to avoid serious monetary losses, to use porcelain or silica crucibles whenever these will give satisfactory service. When platinum utensils are used the danger of theft should always be kept in mind.

PREPARATION OF CRUCIBLES FOR USE

All crucibles, of whatever material, must always be cleaned, ignited and allowed to cool in a desiccator before weighing, since all bodies exposed to the air condense on their surfaces a layer of moisture which increases their weight. The amount and weight of this moisture varies with the humidity of the atmosphere, and the latter may change from hour to hour. The air in the desiccator (see above) is kept at a constant and low humidity by the drying agent which it contains. Bodies which remain in a desiccator for a sufficient time (usually 20-30 minutes) retain, therefore, on their surfaces a constant weight of moisture which is the same day after day, thus insuring constant conditions.

Hot objects, such as ignited crucibles, should be allowed to cool in the air until, when held near the skin, but little heat is noticeable. If this precaution is not taken, the air within the desiccator is strongly heated and expands before the desiccator is covered. As the temperature falls, the air contracts, causing a reduction of air pressure within the covered vessel. When the cover is removed (which is often rendered difficult) the inrush of air from the outside may sweep light particles out of a crucible, thus ruining an entire analysis.

Constant heating of platinum causes a slight crystallization of the surface which, if not removed, penetrates into the crucible. Gentle polishing of the surface destroys the crystalline structure and prevents further damage. If sea sand is used for this purpose, great care is necessary to keep it from the desk, since beakers are easily scratched by it, and subsequently crack on heating.

Platinum crucibles stained in use may often be cleaned by the fusion in them of potassium or sodium acid sulphate, or by heating with ammonium chloride. If the former is used, care should be taken not to heat so strongly as to expel all of the sulphuric acid, since the normal sulphates sometimes expand so rapidly on cooling as to split the crucible. The fused material should be poured out, while hot, on to a !dry! tile or iron surface.

IGNITION OF PRECIPITATES

Most precipitates may, if proper precautions are taken, be ignited without previous drying. If, however, such precipitates can be dried without loss of time to the analyst (as, for example, over night), it is well to submit them to this process. It should, nevertheless, be remembered that a partially dried precipitate often requires more care during ignition than a thoroughly moist one.

The details of the ignition of precipitates vary so much with the character of the precipitate, its moisture content, and temperature to which it is to be heated, that these details will be given under the various procedures which follow.

DETERMINATION OF CHLORINE IN SODIUM CHLORIDE

!Method A. With the Use of a Gooch Filter!

PROCEDURE.—Carefully clean a weighing-tube containing the sodium chloride, handling it as little as possible with the moist fingers, and weigh it accurately to 0.0001 gram, recording the weight at once in the notebook (see Appendix). Hold the tube over the top of a beaker (200-300 cc.), and cautiously remove the stopper, noting carefully that no particles fall from it, or from the tube, elsewhere than into the beaker. Pour out a small portion of the chloride, replace the stopper, and determine by approximate weighing how much has been removed. Continue this procedure until 0.25-0.30 gram has been taken from the tube, then weigh accurately and record the weight beneath the first in the notebook. The difference of the two weights represents the weight of the chloride taken for analysis. Again weigh a second portion of 0.25-0.30 gram into a second beaker of the same size as the first. The beakers should be plainly marked to correspond with the entries in the notebook. Dissolve each portion of the chloride in 150 cc. of distilled water and add about ten drops of dilute nitric acid (sp. gr. 1.20) (Note 2). Calculate the volume of silver nitrate solution required to effect complete precipitation in each case, and add slowly about 5 cc. in excess of that amount, with constant stirring. Heat the solutions cautiously to boiling, stirring occasionally, and continue the heating and stirring until the precipitates settle promptly, leaving a nearly clear supernatant liquid (Note 3). This heating should not take place in direct sunlight (Note 4). The beaker should be covered with a watch-glass, and both boiling and stirring so regulated as to preclude any possibility of loss of material. Add to the clear liquid one or two drops of silver nitrate solution, to make sure that an excess of the reagent is present. If a precipitate, or cloudiness, appears as the drops fall into the solution, heat again, and stir until the whole precipitate has coagulated. The solution is then ready for filtration.

Prepare a Gooch filter as follows: Fold over the top of a Gooch funnel (Fig. 2) a piece of rubber-band tubing, such as is known as "bill-tie" tubing, and fit into the mouth of the funnel a perforated porcelain crucible (Gooch crucible), making sure that when the crucible is gently forced into the mouth of the funnel an airtight joint results. (A small 1 or 1-1/4-inch glass funnel may be used, in which case the rubber tubing is stretched over the top of the funnel and then drawn up over the side of the crucible until an air-tight joint is secured.)

[ILLUSTRATION: FIG. 2]

Fit the funnel into the stopper of a filter bottle, and connect the filter bottle with the suction pump. Suspend some finely divided asbestos, which has been washed with acid, in 20 to 30 cc. of water (Note 1); allow this to settle, pour off the very fine particles, and then pour some of the mixture cautiously into the crucible until an even felt of asbestos, not over 1/32 inch in thickness, is formed. A gentle suction must be applied while preparing this felt. Wash the felt thoroughly by passing through it distilled water until all fine or loose particles are removed, increasing the suction at the last until no more water can be drawn out of it; place on top of the felt the small, perforated porcelain disc and hold it in place by pouring a very thin layer of asbestos over it, washing the whole carefully; then place the crucible in a small beaker, and place both in a drying closet at 100-110°C. for thirty to forty minutes. Cool the crucible in a desiccator, and weigh. Heat again for twenty to thirty minutes, cool, and again weigh, repeating this until the weight is constant within 0.0003 gram. The filter is then ready for use.

Place the crucible in the funnel, and apply a gentle suction, !after which! the solution to be filtered may be poured in without disturbing the asbestos felt. When pouring liquid onto a Gooch filter hold the stirring-rod at first well down in the crucible, so that the liquid does not fall with any force upon the asbestos, and afterward keep the crucible will filled with the solution.

Pour the liquid above the silver chloride slowly onto the filter, leaving the precipitate in the beaker as far as possible. Wash the precipitate twice by decantation with warm water; then transfer it to the filter with the aid of a stirring-rod with a rubber tip and a stream from the wash-bottle.

Examine the first portions of the filtrate which pass through the filter with great care for asbestos fibers, which are most likely to be lost at this point. Refilter the liquid if any fibers are visible. Finally, wash the precipitate thoroughly with warm water until free from soluble silver salts. To test the washings, disconnect the suction at the flask and remove the funnel or filter tube from the suction flask. Hold the end of the tube over the mouth of a small test tube and add from a wash-bottle 2-3 cc. of water. Allow the water to drip through into the test tube and add a drop of dilute hydrochloric acid. No precipitate or cloud should form in the wash-water (Note 16). Dry the filter and contents at 100-110°C. until the weight is constant within 0.0003 gram, as described for the preparation of the filter. Deduct the weight of the dry crucible from the final weight, and from the weight of silver chloride thus obtained calculate the percentage of chlorine in the sample of sodium chloride.

[Note 1: The washed asbestos for this type of filter is prepared by digesting in concentrated hydrochloric acid, long-fibered asbestos which has been cut in pieces of about 0.5 cm. in length. After digestion, the asbestos is filtered off on a filter plate and washed with hot, distilled water until free from chlorides. A small portion of the asbestos is shaken with water, forming a thin suspension, which is bottled and kept for use.]

[Note 2: The nitric acid is added before precipitation to lessen the tendency of the silver chloride to carry down with it other substances which might be precipitated from a neutral solution. A large excess of the acid would exert a slight solvent action upon the chloride.]

[Note 3: The solution should not be boiled after the addition of the nitric acid before the presence of an excess of silver nitrate is assured, since a slight interaction between the nitric acid and the sodium chloride is possible, by which a loss of chlorine, either as such or as hydrochloric acid, might ensue. The presence of an excess of the precipitant can usually be recognized at the time of its addition, by the increased readiness with which the precipitate coagulates and settles.]

[Note 4: The precipitate should not be exposed to strong sunlight, since under those conditions a reduction of the silver chloride ensues which is accompanied by a loss of chlorine. The superficial alteration which the chloride undergoes in diffused daylight is not sufficient to materially affect the accuracy of the determination. It should be noted, however, that a slight error does result from the effect of light upon the silver chloride precipitate and in cases in which the greatest obtainable accuracy is required, the procedure described under "Method B" should be followed, in which this slight reduction of the silver chloride is corrected by subsequent treatment with nitric and hydrochloric acids.]

[Note 5: The asbestos used in the Gooch filter should be of the finest quality and capable of division into minute fibrous particles. A coarse felt is not satisfactory.]

[Note 6: The precipitate must be washed with warm water until it is absolutely free from silver and sodium nitrates. It may be assumed that the sodium salt is completely removed when the wash-water shows no evidence of silver. It must be borne in mind that silver chloride is somewhat soluble in hydrochloric acid, and only a single drop should be added. The washing should be continued until no cloudiness whatever can be detected in 3 cc. of the washings.

Silver chloride is but slightly soluble in water. The solubility varies with its physical condition within small limits, and is about 0.0018 gram per liter at 18°C. for the curdy variety usually precipitated. The chloride is also somewhat soluble in solutions of many chlorides, in solutions of silver nitrate, and in concentrated nitric acid.

As a matter of economy, the filtrate, which contains whatever silver nitrate was added in excess, may be set aside. The silver can be precipitated as chloride and later converted into silver nitrate.]

[Note 7: The use of the Gooch filter commends itself strongly when a considerable number of halogen determinations are to be made, since successive portions of the silver halides may be filtered on the same filter, without the removal of the preceding portions, until the crucible is about two thirds filled. If the felt is properly prepared, filtration and washing are rapidly accomplished on this filter, and this, combined with the possibility of collecting several precipitates on the same filter, is a strong argument in favor of its use with any but gelatinous precipitates.]

!Method B. With the Use of a Paper Filter!

PROCEDURE.—Weigh out two portions of sodium chloride of about 0.25-0.3 gram each and proceed with the precipitation of the silver chloride as described under Method A above. When the chloride is ready for filtration prepare two 9 cm. washed paper filters (see Appendix). Pour the liquid above the precipitates through the filters, wash twice by decantation and transfer the precipitates to the filters, finally washing them until free from silver solution as described. The funnel should then be covered with a moistened filter paper by stretching it over the top and edges, to which it will adhere on drying. It should be properly labeled with the student's name and desk number, and then placed in a drying closet, at a temperature of about 100-110°C., until completely dry.

The perfectly dry filter is then opened over a circular piece of clean, smooth, glazed paper about six inches in diameter, placed upon a larger piece about twelve inches in diameter. The precipitate is removed from the filter as completely as possible by rubbing the sides gently together, or by scraping them cautiously with a feather which has been cut close to the quill and is slightly stiff (Note 1). In either case, care must be taken not to rub off any considerable quantity of the paper, nor to lose silver chloride in the form of dust. Cover the precipitate on the glazed paper with a watch-glass to prevent loss of fine particles and to protect it from dust from the air. Fold the filter paper carefully, roll it into a small cone, and wind loosely around !the top! a piece of small platinum wire (Note 2). Hold the filter by the wire over a small porcelain crucible (which has been cleaned, ignited, cooled in a desiccator, and weighed), ignite it, and allow the ash to fall into the crucible. Place the crucible upon a clean clay triangle, on its side, and ignite, with a low flame well at its base, until all the carbon of the filter has been consumed. Allow the crucible to cool, add two drops of concentrated nitric acid and one drop of concentrated hydrochloric acid, and heat !very cautiously!, to avoid spattering, until the acids have been expelled; then transfer the main portion of the precipitate from the glazed paper to the cooled crucible, placing the latter on the larger piece of glazed paper and brushing the precipitate from the smaller piece into it, sweeping off all particles belonging to the determination.

Moisten the precipitate with two drops of concentrated nitric acid and one drop of concentrated hydrochloric acid, and again heat with great caution until the acids are expelled and the precipitate is white, when the temperature is slowly raised until the silver chloride just begins to fuse at the edges (Note 3). The crucible is then cooled in a desiccator and weighed, after which the heating (without the addition of acids) is repeated, and it is again weighed. This must be continued until the weight is constant within 0.0003 gram in two consecutive weighings. Deduct the weight of the crucible, and calculate the percentage of chlorine in the sample of sodium chloride taken for analysis.

[Note 1: The separation of the silver chloride from the filter is essential, since the burning carbon of the paper would reduce a considerable quantity of the precipitate to metallic silver, and its complete reconversion to the chloride within the crucible, by means of acids, would be accompanied by some difficulty. The small amount of silver reduced from the chloride adhering to the filter paper after separating the bulk of the precipitate, and igniting the paper as prescribed, can be dissolved in nitric acid, and completely reconverted to chloride by hydrochloric acid. The subsequent addition of the two acids to the main portion of the precipitate restores the chlorine to any chloride which may have been partially reduced by the sunlight. The excess of the acids is volatilized by heating.]

[Note 2: The platinum wire is wrapped around the top of the filter during its incineration to avoid contact with any reduced silver from the reduction of the precipitate. If the wire were placed nearer the apex, such contact could hardly be avoided.]

[Note 3: Silver chloride should not be heated to complete fusion, since a slight loss by volatilization is possible at high temperatures. The temperature of fusion is not always sufficient to destroy filter shreds; hence these should not be allowed to contaminate the precipitate.]

DETERMINATION OF IRON AND OF SULPHUR IN FERROUS AMMONIUM SULPHATE,

$FESO_4 \cdot (NH_4)_2 SO_4 \cdot 6H_2O$

DETERMINATION OF IRON

PROCEDURE.—Weigh out into beakers (200-250 cc.) two portions of the sample (Note 1) of about 1 gram each and dissolve these in 50 cc. of water, to which 1 cc. of dilute hydrochloric acid (sp. gr. 1.12) has been added (Note 2). Heat the solution to boiling, and while at the boiling point add concentrated nitric acid (sp. gr. 1.42), !drop by drop! (noting the volume used), until the brown coloration, which appears after the addition of a part of the nitric acid, gives place to a yellow or red (Note 3). Avoid a large excess of nitric acid, but be sure that the action is complete. Pour this solution cautiously into about 200 cc. of water, containing a slight excess of ammonia. Calculate for this purpose the amount of aqueous ammonia required to neutralize the hydrochloric and nitric acids added (see Appendix for data), and also to precipitate the iron as ferric hydroxide from the weight of the ferrous ammonium sulphate taken for analysis, assuming it to be pure (Note 4). The volume thus calculated will be in excess of that actually required for precipitation, since the acids are in part consumed in the oxidation process, or are volatilized. Heat the solution to boiling, and allow the precipitated ferric hydroxide to settle. Decant the clear liquid through a washed filter (9 cm.), keeping as much of the precipitate in the beaker as possible. Wash twice by decantation with 100 cc. of hot water. Reserve the filtrate. Dissolve the iron from the filter with hot, dilute hydrochloric acid (sp. gr. 1.12), adding it in small portions, using as little as possible and noting the volume used. Collect the solution in the beaker in which precipitation took place. Add 1 cc. of nitric acid (sp. gr. 1.42), boil for a few moments, and again pour into a calculated excess of ammonia.

Wash the precipitate twice by decantation, and finally transfer it to the original filter. Wash continuously with hot water until finally 3 cc. of the washings, acidified with nitric acid (Note 5), show no evidences of the presence of chlorides when tested with silver nitrate. The filtrate and washings are combined with those from the first precipitation and treated for the determination of sulphur, as prescribed on page 112.

[Note 1: If a selection of pure material for analysis is to be made, crystals which are cloudy are to be avoided on account of loss of water of crystallization; and also those which are red, indicating the presence of ferric iron. If, on the other hand, the value of an average sample of material is desired, it is preferable to grind the whole together, mix thoroughly, and take a sample from the mixture for analysis.]

[Note 2: When aqueous solutions of ferrous compounds are heated in the air, oxidation of the Fe^{++} ions to Fe^{+++} ions readily occurs in the absence of free acid. The H^{+} and OH^{-} ions from water are involved in the oxidation process and the result is, in effect, the formation of some ferric hydroxide which tends to separate. Moreover, at the boiling temperature, the ferric sulphate produced by the oxidation hydrolyzes in part with the formation of a basic ferric sulphate, which also tends to separate from solution. The addition of the hydrochloric acid prevents the formation of ferric hydroxide, and so far reduces the ionization of the water that the hydrolysis of the ferric sulphate is also prevented, and no precipitation occurs on heating.]

[Note 3: The nitric acid, after attaining a moderate strength, oxidizes the Fe^{++} ions to Fe^{+++} ions with the formation of an intermediate nitroso-compound similar in character to that formed in the "ring-test" for nitrates. The nitric oxide is driven out by heat, and the solution then shows by its color the presence of ferric compounds. A drop of the oxidized solution should be tested on a watch-glass with potassium ferricyanide, to insure a complete oxidation. This oxidation of the iron is necessary, since Fe^{++} ions are not completely precipitated by ammonia.

The ionic changes which are involved in this oxidation are perhaps most simply expressed by the equation

$$3Fe^{++} + NO_3^- + 4H^+ \longrightarrow 3Fe^{+++} + 2H_2O + NO,$$

the H^+ ions coming from the acid in the solution, in this case either the nitric or the hydrochloric acid. The full equation on which this is based may be written thus:

$$6FeSO_4 + 2HNO_3 + 6HCl \longrightarrow 2Fe_2(SO_4)_3 + 2FeCl_3 + 2NO + 4H_2O,$$

assuming that only enough nitric acid is added to complete the oxidation.]

[Note 4: The ferric hydroxide precipitate tends to carry down some sulphuric acid in the form of basic ferric sulphate. This tendency is lessened if the solution of the iron is added to an excess of OH^- ions from the ammonium hydroxide, since under these conditions immediate and complete precipitation of the ferric hydroxide ensues. A gradual neutralization with ammonia would result in the local formation of a neutral solution within the liquid, and subsequent deposition of a basic sulphate as a consequence of a local deficiency of OH^- ions from the NH_4OH and a partial hydrolysis of the ferric salt. Even with this precaution the entire absence of sulphates from the first iron precipitate is not assured. It is, therefore, redissolved and again thrown down by ammonia. The organic matter of the filter paper may occasion a partial reduction of the iron during solution, with consequent possibility of incomplete subsequent precipitation with ammonia. The nitric acid is added to reoxidize this iron.

To avoid errors arising from the solvent action of ammoniacal liquids upon glass, the iron precipitate should be filtered without unnecessary delay.]

[Note 5: The washings from the ferric hydroxide are acidified with nitric acid, before testing with silver nitrate, to destroy the ammonia which is a solvent of silver chloride.

The use of suction to promote filtration and washing is permissible, though not prescribed. The precipitate should not be allowed to dry during the washing.]

!Ignition of the Iron Precipitate!

Heat a platinum or porcelain crucible, cool it in a desiccator and weigh, repeating until a constant weight is obtained.

Fold the top of the filter paper over the moist precipitate of ferric hydroxide and transfer it cautiously to the crucible. Wipe the inside of the funnel with a small fragment of washed filter paper, if necessary, and place the paper in the crucible.

Incline the crucible on its side, on a triangle supported on a ringstand, and stand the cover on edge at the mouth of the crucible. Place a burner below the front edge of the crucible, using a low flame and protecting it from drafts of air by means of a chimney. The heat from the burner is thus reflected into the crucible and dries the precipitate without danger of loss as the result of a sudden generation of steam within the mass of ferric hydroxide. As the drying progresses the burner may be gradually moved toward the base of the crucible and the flame increased until the paper of the filter begins to char and finally to smoke, as the volatile matter is expelled. This is known as "smoking off" a filter, and the temperature should not be raised sufficiently high during this process to cause the paper to ignite, as the air currents produced by the flame of the blazing paper may carry away particles of the precipitate.

When the paper is fully charred, move the burner to the base of the crucible and raise the temperature to the full heat of the burner for fifteen minutes, with the crucible still inclined on its side, but without the cover (Note 1). Finally set the crucible upright in the triangle, cover it, and heat at the full temperature of a blast lamp or other high temperature burner. Cool and weigh in the usual manner (Note 2). Repeat the strong heating until the weight is constant within 0.0003 gram.

From the weight of ferric oxide (Fe_2O_3) calculate the percentage of iron (Fe) in the sample (Note 3).

[Note 1: These directions for the ignition of the precipitate must be closely followed. A ready access of atmospheric oxygen is of special importance to insure the reoxidation to ferric oxide of any iron which may be reduced to magnetic oxide (Fe_3O_4) during the combustion of the filter. The final heating over the blast lamp is essential for the complete expulsion of the last traces of water from the hydroxide.]

[Note 2: Ignited ferric oxide is somewhat hygroscopic. On this account the weighings must be promptly completed after removal from the desiccator. In all weighings after the first it is well to place the weights upon the balance-pan before removing the crucible from the desiccator. It is then only necessary to move the rider to obtain the weight.]

[Note 3: The gravimetric determination of aluminium or chromium is comparable with that of iron just described, with the additional precaution that the solution must be boiled until it contains but a very slight excess of ammonia, since the hydroxides of aluminium and chromium are more soluble than ferric hydroxide.

The most important properties of these hydroxides, from a quantitative standpoint, other than those mentioned, are the following: All are precipitable by the hydroxides of sodium and potassium, but always inclose some of the precipitant, and should be reprecipitated with ammonium hydroxide before ignition to oxides. Chromium and aluminium hydroxides dissolve in an excess of the caustic alkalies and form anions, probably of the formula AlO_2^{-} and CrO_2^{-}. Chromium hydroxide is reprecipitated from this solution on boiling. When first precipitated the hydroxides are all readily soluble in acids, but aluminium hydroxide dissolves with considerable difficulty after standing or boiling for some time. The precipitation of the hydroxides is promoted by the presence of ammonium chloride, but is partially or entirely prevented by the presence of tartaric or citric acids, glycerine, sugars, and some other forms of soluble organic matter. The hydroxides yield on ignition an oxide suitable for weighing (Al_2O_3, Cr_2O_3, Fe_2O_3).]

DETERMINATION OF SULPHUR

PROCEDURE.—Add to the combined filtrates from the ferric hydroxide about 0.6 gram of anhydrous sodium carbonate; cover the beaker, and then add dilute hydrochloric acid (sp. gr. 1.12) in moderate excess and evaporate to dryness on the water bath. Add 10 cc. of concentrated hydrochloric acid (sp. gr. 1.20) to the residue, and again evaporate to dryness on the bath. Dissolve the residue in water, filter if not clear, transfer to a 700 cc. beaker, dilute to about 400 cc., and cautiously add hydrochloric acid until the solution shows a distinctly acid reaction (Note 1). Heat the solution to boiling, and add !very slowly! and with constant stirring, 20 cc. in excess of the calculated amount of a hot barium chloride solution, containing about 20 grams $BaCl_2 \cdot 2H_2O$ per liter (Notes 2 and 3). Continue the boiling for about two minutes, allow the precipitate to settle, and decant the liquid at the end of half an hour (Note 4). Replace the beaker containing the original filtrate by a clean beaker, wash the precipitated sulphate by decantation with hot water, and subsequently upon the filter until it is freed from chlorides, testing the washings as described in the determination of iron. The filter is then transferred to a platinum or porcelain crucible and ignited, as described above, until the weight is constant (Note 5). From the weight of barium sulphate ($BaSO_4$) obtained, calculate the percentage of sulphur (S) in the sample.

[Note 1: Barium sulphate is slightly soluble in hydrochloric acid, even dilute, probably as a result of the reduction in the degree of dissociation of sulphuric acid in the presence of the H^+ ions of the hydrochloric acid, and possibly because of the formation of a complex anion made up of barium and chlorine; hence only the smallest excess should be added over the amount required to acidify the solution.]

[Note 2: The ionic changes involved in the precipitation of barium sulphate are very simple:

$Ba^{++} + SO_4^{-} \longrightarrow [BaSO_4]$

This case affords one of the best illustrations of the effect of an excess of a precipitant in decreasing the solubility of a precipitate. If the conditions are considered which exist at the moment when just enough of the Ba^{++} ions have been added to correspond to the SO_4^{-} ions in the solution, it will be seen that nearly all of the barium sulphate has been precipitated, and that the small amount which then remains in the solution which is in contact with the precipitate must represent a saturated solution for the existing temperature, and that this solution is comparable with a solution of sugar to which more sugar has been added than will dissolve. It should be borne in mind that the quantity of barium sulphate in this !saturated solution is a constant quantity! for the existing conditions. The dissolved barium sulphate, like any electrolyte, is dissociated, and the equilibrium conditions may be expressed thus:

(!Conc'n Ba^{++} x Conc'n SO_4^{-}!)/(Conc'n $BaSO_4$) = Const.!,

and since !Conc'n $BaSO_4$! for the saturated solution has a constant value (which is very small), it may be eliminated, when the expression becomes !Conc'n Ba^{++} x Conc'n SO_4^{-} = Const.!, which is the "solubility product" of $BaSO_4$. If, now, an excess of the precipitant, a soluble barium salt, is added in the form of a relatively concentrated solution (the slight change of volume of a few cubic centimeters may be disregarded for the present discussion) the concentration of the Ba^{++} ions is much increased, and as a consequence the !Conc'n SO_4! must decrease in proportion if the value of the expression is to remain constant, which is a requisite condition if the law of mass action upon which our argument depends holds true. In other words, SO_4^{-} ions must combine with some of the added Ba^{++} ions to form $[BaSO_4]$; but it will be recalled that the solution is already saturated with $BaSO_4$, and this freshly formed quantity must, therefore, separate and add itself to the precipitate. This is exactly what is desired in order to insure more complete precipitation and greater accuracy, and leads to the conclusion that the larger the excess of the precipitant added the more successful the analysis; but a practical limit is placed upon the quantity of the precipitant which may be properly added by other conditions, as stated in the following note.]

[Note 3: Barium sulphate, in a larger measure than most compounds, tends to carry down other substances which are present in the solution from which it separates, even when these other substances are relatively soluble, and including the barium chloride used as the precipitant. This is also notably true in the case of nitrates and chlorates of the alkalies, and of ferric compounds; and, since in this analysis ammonium nitrate has resulted from the neutralization of the excess of the nitric acid added to oxidize the iron, it is essential that this should be destroyed by repeated evaporation with a relatively large quantity of hydrochloric acid. During evaporation a mutual decomposition of the two acids takes place, and the nitric acid is finally decomposed and expelled by the excess of hydrochloric acid.

Iron is usually found in the precipitate of barium sulphate when thrown down from hot solutions in the presence of ferric salts. This, according to Kuster and Thiel (!Zeit. anorg. Chem.!, 22, 424), is due to the formation of a complex ion $(Fe(SO_4)_2)$ which precipitates with the Ba^{++} ion, while Richards (!Zeit. anorg. Chem.!, 23, 383) ascribes it to hydrolytic action, which causes the formation of a basic ferric complex which is occluded in the barium precipitate. Whatever the character of the compound may be, it has been shown that it loses sulphuric anhydride upon ignition, causing low results, even though the precipitate contains iron.

The contamination of the barium sulphate by iron is much less in the presence of ferrous than ferric salts. If, therefore, the sulphur alone were to be determined in the ferrous ammonium sulphate, the precipitation by barium might be made directly from an aqueous solution of the salt, which had been made slightly acid with hydrochloric acid.]

[Note 4: The precipitation of the barium sulphate is probably complete at the end of a half-hour, and the solution may safely be filtered at the expiration of that time if it is desired to hasten the analysis.

As already noted, many precipitates of the general character of this sulphate tend to grow more coarsely granular if digested for some time with the liquid from which they have separated. It is therefore well to allow the precipitate to stand in a warm place for several hours, if practicable, to promote ease of filtration. The filtrate and washings should always be carefully examined for minute quantities of the sulphate which may pass through the pores of the filter. This is best accomplished by imparting to the filtrate a gentle rotary motion, when the sulphate, if present, will collect at the center of the bottom of the beaker.]

[Note 5: A reduction of barium sulphate to the sulphide may very readily be caused by the reducing action of the burning carbon of the filter, and much care should be taken to prevent any considerable reduction from this cause. Subsequent ignition, with ready access of air, reconverts the sulphide to sulphate unless a considerable reduction has occurred. In the latter case it is expedient to add one or two drops of sulphuric acid and to heat cautiously until the excess of acid is expelled.]

[Note 6: Barium sulphate requires about 400,000 parts of water for its solution. It is not decomposed at a red heat but suffers loss, probably of sulphur trioxide, at a temperature above 900°C.]

DETERMINATION OF SULPHUR IN BARIUM SULPHATE

PROCEDURE.—Weigh out, into platinum crucibles, two portions of about 0.5 gram of the sulphate. Mix each in the crucible with five to six times its weight of anhydrous sodium carbonate. This can best be done by placing the crucible on a piece of glazed paper and stirring the mixture with a clean, dry stirring-rod, which may finally be wiped off with a small fragment of filter paper, the latter being placed in the crucible. Cover the crucible and heat until a quiet, liquid fusion ensues. Remove the burner, and tip the crucible until the fused mass flows nearly to its mouth. Hold it in that position until the mass has solidified. When cold, the material may usually be detached in a lump by tapping the crucible or gently pressing it near its upper edge. If it still adheres, a cubic centimeter or so of water may be placed in the cold crucible and cautiously brought to boiling, when the cake will become loosened and may be removed and placed in about 250 cc. of hot, distilled water to dissolve. Clean the crucible completely, rubbing the sides with a rubber-covered stirring-rod, if need be.

When the fused mass has completely disintegrated and nothing further will dissolve, decant the solution from the residue of barium carbonate (Note 1). Pour over the residue 20 cc. of a solution of sodium carbonate and 10 cc. of water and heat to gentle boiling for about three minutes (Note 2). Filter off the carbonate and wash it with hot water, testing the slightly acidified washings for sulphate and preserving any precipitates which appear in these tests. Acidify the filtrate with hydrochloric acid until just acid, bring to boiling, and slowly add hot barium chloride solution, as in the preceding determination. Add also any tests from the washings in which precipitates have appeared. Filter, wash, ignite, and weigh.

From the weight of barium sulphate, calculate the percentage of sulphur (S) in the sample.

[Note 1: This alkaline fusion is much employed to disintegrate substances ordinarily insoluble in acids into two components, one of which is water soluble and the other acid soluble. The reaction involved is:

$$BaSO_4 + Na_2CO_3 \longrightarrow BaCO_3 + Na_2SO_4.$$

As the sodium sulphate is soluble in water, and the barium carbonate insoluble, a separation between them is possible and the sulphur can be determined in the water-soluble portion.

It should be noted that this method can be applied to the purification of a precipitate of barium sulphate if contaminated by most of the substances mentioned in Note 3 on page 114. The impurities pass into the water solution together with the sodium sulphate, but, being present in such minute amounts, do not again precipitate with the barium sulphate.]

[Note 2: The barium carbonate is boiled with sodium carbonate solution before filtration because the reaction above is reversible; and it is only by keeping the sodium carbonate present in excess until nearly all of the sodium sulphate solution has been removed by filtration that the reversion of some of the barium carbonate to barium sulphate is prevented. This is an application of the principle of mass action, in which the concentration of the reagent (the carbonate ion) is kept as high as practicable and that of the sulphate ion as low as possible, in order to force the reaction in the desired direction (see Appendix).]

DETERMINATION OF PHOSPHORIC ANHYDRIDE IN APATITE

The mineral apatite is composed of calcium phosphate, associated with calcium chloride, or fluoride. Specimens are easily obtainable which are nearly pure and leave on treatment with acid only a slight siliceous residue.

For the purpose of gravimetric determination, phosphoric acid is usually precipitated from ammoniacal solutions in the form of magnesium ammonium phosphate which, on ignition, is converted into magnesium pyrophosphate. Since the calcium phosphate of the apatite is also insoluble in ammoniacal solutions, this procedure cannot be applied directly. The separation of the phosphoric acid from the calcium must first be accomplished by precipitation in the form of ammonium phosphomolybdate in nitric acid solution, using ammonium molybdate as the precipitant. The "yellow precipitate," as it is often called, is not always of a definite composition, and therefore not suitable for direct weighing, but may be dissolved in ammonia, and the phosphoric acid thrown out as magnesium ammonium phosphate from the solution.

Of the substances likely to occur in apatite, silicic acid alone interferes with the precipitation of the phosphoric acid in nitric acid solution.

PRECIPITATION OF AMMONIUM PHOSPHOMOLYBDATE

PROCEDURE.—Grind the mineral in an agate mortar until no grit is perceptible. Transfer the substance to a weighing-tube, and weigh out two portions, not exceeding 0.20 gram each (Note 1) into two beakers of about 200 cc. capacity. Pour over them 20 cc. of dilute nitric acid (sp. gr. 1.2) and warm gently until solvent action has apparently ceased. Evaporate the solution cautiously to dryness, heat the residue for about an hour at 100-110°C., and treat it again with nitric acid as described above; separate the residue of silica by filtration on a small filter (7 cm.) and wash with warm water, using as little as possible (Note 2). Receive the filtrate in a beaker (200-500 cc.). Test the washings with ammonia for calcium phosphate, but add all such tests in which a precipitate appears to the original nitrate (Note 3). The filtrate and washings must be kept as small as possible and should not exceed 100 cc. in volume. Add aqueous ammonia (sp. gr. 0.96) until the precipitate of calcium phosphate first produced just fails to redissolve, and then add a few drops of nitric acid until this is again brought into solution (Note 4). Warm the solution until it cannot be comfortably held in the hand (about 60°C.) and, after removal of the burner, add 75 cc. of ammonium molybdate solution which has been !gently! warmed, but which must be perfectly clear. Allow the mixture to stand at a temperature of about 50 or 60°C. for twelve hours (Notes 5 and 6). Filter off the yellow precipitate on a 9 cm. filter, and wash by decantation with a solution of ammonium nitrate made acid with nitric acid.[1] Allow the precipitate to remain in the beaker as far as possible. Test the washings for calcium with ammonia and ammonium oxalate (Note 3).

[Footnote 1: This solution is prepared as follows: Mix 100 cc. of ammonia solution (sp. gr. 0.96) with 325 cc. of nitric acid (sp. gr. 1.2) and dilute with 100 cc. of water.]

Add 10 cc. of molybdate solution to the nitrate, and leave it for a few hours. It should then be carefully examined for a !yellow! precipitate; a white precipitate may be neglected.

[Note 1: Magnesium ammonium phosphate, as noted below, is slightly soluble under the conditions of operation. Consequently the unavoidable errors of analysis are greater in this determination than in those which have preceded it, and some divergence may be expected in duplicate analyses. It is obvious that the larger the amount of substance taken for analysis the less will be the relative loss or gain due to unavoidable experimental errors; but, in this instance, a check is placed upon the amount of material which may be taken both by the bulk of the resulting precipitate of ammonium phosphomolybdate and by the excessive amount of ammonium molybdate required to effect complete separation of the phosphoric acid, since a liberal excess above the theoretical quantity is demanded. Molybdic acid is one of the more expensive reagents.]

[Note 2: Soluble silicic acid would, if present, partially separate with the phosphomolybdate, although not in combination with molybdenum. Its previous removal by dehydration is therefore necessary.]

[Note 3: When washing the siliceous residue the filtrate may be tested for calcium by adding ammonia, since that reagent neutralizes the acid which holds the calcium phosphate in solution and causes precipitation; but after the removal of the phosphoric acid in combination with the molybdenum, the addition of an oxalate is required to show the presence of calcium.]

[Note 4: An excess of nitric acid exerts a slight solvent action, while ammonium nitrate lessens the solubility; hence the neutralization of the former by ammonia.]

[Note 5: The precipitation of the phosphomolybdate takes place more promptly in warm than in cold solutions, but the temperature should not exceed 60°C. during precipitation; a higher temperature tends to separate molybdic acid from the solution. This acid is nearly white, and its deposition in the filtrate on long standing should not be mistaken for a second precipitation of the yellow precipitate. The addition of 75 cc. of ammonium molybdate solution insures the presence of a liberal excess of the reagent, but the filtrate should be tested as in all quantitative procedures.

The precipitation is probably complete in many cases in less than twelve hours; but it is better, when practicable, to allow the solution to stand for this length of time. Vigorous shaking or stirring promotes the separation of the precipitate.]

[Note 6: The composition of the "yellow precipitate" undoubtedly varies slightly with varying conditions at the time of its formation. Its composition may probably fairly be represented by the formula, $(NH_4)_3PO_4 \cdot 12MoO_3 \cdot H_2O$, when precipitated under the conditions prescribed in the procedure. Whatever other variations may occur in its composition, the ratio of 12 MoO_3:1 P seems to hold, and this fact is utilized in volumetric processes for the determination of phosphorus, in which the molybdenum is reduced to a lower oxide and reoxidized by a standard solution of potassium permanganate. In principle, the procedure is comparable with that described for the determination of iron by permanganate.]

PRECIPITATION OF MAGNESIUM AMMONIUM PHOSPHATE

PROCEDURE. — Dissolve the precipitate of phosphomolybdate upon the filter by pouring through it dilute aqueous ammonia (one volume of dilute ammonia (sp. gr. 0.96) and three volumes of water, which should be carefully measured), and receive the solution in the beaker containing the bulk of the precipitate. The total volume of nitrate and washings should not much exceed 100 cc. Acidify the solution with dilute hydrochloric acid, and heat it nearly to boiling. Calculate the volume of magnesium ammonium chloride solution ("magnesia mixture") required to precipitate the phosphoric acid, assuming 40 per cent P_2O_5 in the apatite. Measure out about 5 cc. in excess of this amount, and pour it into the acid solution. Then add slowly dilute ammonium hydroxide (1 volume of strong ammonia (sp. gr. 0.90) and 9 volumes of water), stirring constantly until a precipitate forms. Then add a volume of filtered, concentrated ammonia (sp. gr. 0.90) equal to one third of the volume of liquid in the beaker (Note 1). Allow the whole to cool. The precipitated magnesium ammonium phosphate should then be definitely crystalline in appearance (Note 2). (If it is desired to hasten the precipitation, the solution may be cooled, first in cold and then in ice-water, and stirred !constantly! for half an hour, when precipitation will usually be complete.)

Decant the clear liquid through a filter, and transfer the precipitate to the filter, using as wash-water a mixture of one volume of concentrated ammonia and three volumes of water. It is not necessary to clean the beaker completely or to wash the precipitate thoroughly at this point, as it is necessary to purify it by reprecipitation.

[Note 1: Magnesium ammonium phosphate is not a wholly insoluble substance, even under the most favorable analytical conditions. It is least soluble in a liquid containing one fourth of its volume of concentrated aqueous ammonia (sp. gr. 0.90) and this proportion should be carefully maintained as prescribed in the procedure. On account of this slight solubility the volume of solutions should be kept as small as possible and the amount of wash-water limited to that absolutely required.

A large excess of the magnesium solution tends both to throw out magnesium hydroxide (shown by a persistently flocculent precipitate) and to cause the phosphate to carry down molybdic acid. The tendency of the magnesium precipitate to carry down molybdic acid is also increased if the solution is too concentrated. The volume should not be less than 90 cc., nor more than 125 cc., at the time of the first precipitation with the magnesia mixture.]

[Note 2: The magnesium ammonium phosphate should be perfectly crystalline, and will be so if the directions are followed. The slow addition of the reagent is essential, and the stirring not less so. Stirring promotes the separation of the precipitate and the formation of larger crystals, and may therefore be substituted for digestion in the cold. The stirring-rod must not be allowed to scratch the glass, as the crystals adhere to such scratches and are removed with difficulty.]

REPRECIPITATION AND IGNITION OF MAGNESIUM AMMONIUM PHOSPHATE

A single precipitation of the magnesium compound in the presence of molybdenum compounds rarely yields a pure product. The molybdenum can be removed by solution of the precipitate in acid and precipitation of the molybdenum by sulphureted hydrogen, after which the magnesium precipitate may be again thrown down. It is usually more satisfactory to dissolve the magnesium precipitate and reprecipitate the phosphate as magnesium ammonium phosphate as described below.

PROCEDURE. — Dissolve the precipitate from the filter in a little dilute hydrochloric acid (sp. gr. 1.12), allowing the acid solution to run into the beaker in which the original precipitation was made (Note 1). Wash the filter with water until the wash-water shows no test for chlorides, but avoid an unnecessary amount of wash-water. Add to the solution 2 cc. (not more) of magnesia mixture, and then dilute ammonium hydroxide solution (sp. gr. 0.96), drop by drop, with constant stirring, until the liquid smells distinctly of ammonia. Stir for a few moments and then add a volume of strong ammonia (sp. gr. 0.90), equal to one third of the volume of the solution. Allow the solution to stand for some hours, and then filter off the magnesium ammonium phosphate, which should be distinctly crystalline in character. Wash the precipitate with dilute ammonia water, as prescribed above, until, finally, 3 cc. of the washings, after acidifying with nitric acid, show no evidence of chlorides. Test both filtrates for complete precipitation by adding a few cubic centimeters of magnesia mixture and allowing them to stand for some time.

Transfer the moist precipitate to a weighed porcelain or platinum crucible and ignite, using great care to raise the temperature slowly while drying the filter in the crucible, and to insure the ready access of oxygen during the combustion of the filter paper, thus guarding against a possible reduction of the phosphate, which would result in disastrous consequences both to the crucible, if of platinum, and the analysis. Do not raise the temperature above moderate redness until the precipitate is white. (Keep this precaution well in mind.) Ignite finally at the highest temperature of the Tirrill burner, and repeat the heating until the weight is constant. If the ignited precipitate is persistently discolored by particles of unburned carbon, moisten the mass with a drop or two of concentrated nitric acid and heat cautiously, finally igniting strongly. The acid will dissolve magnesium pyrophosphate from the surface of the particles of carbon, which will then burn away. Nitric acid also aids as an oxidizing agent in supplying oxygen for the combustion of the carbon.

From the weight of magnesium pyrophosphate ($Mg_2P_2O_7$) obtained, calculate the phosphoric anhydride (P_2O_5) in the sample of apatite.

[Note 1: The ionic change involved in the precipitation of the magnesium compound is

$PO_4^{---} + NH_4^{+} + Mg^{++} \rightarrow [MgNH_4PO_4]$.

The magnesium ammonium phosphate is readily dissolved by acids, even those which are no stronger than acetic acid. This is accounted for by the fact that two of the ions into which phosphoric acid may dissociate, the HPO_4^{-} or $H_2PO_4^{-}$ ions, exhibit the characteristics of very weak acids, in that they show almost no tendency to dissociate further into H^+ and PO_4^- ions. Consequently the ionic changes which occur when the magnesium ammonium phosphate is brought into contact with an acid may be typified by the reaction:

$H^+ + Mg^{++} + NH_4^+ + PO_4^{--} \longrightarrow Mg^{++} + NH_4^+ + HPO_4^-$;

that is, the PO_4^- ions and the H^+ ions lose their identity in the formation of the new ion, HPO_4^-, and this continues until the magnesium ammonium phosphate is entirely dissolved.]

[Note 2: During ignition the magnesium ammonium phosphate loses ammonia and water and is converted into magnesium pyrophosphate:

$2MgNH_4PO_4 \longrightarrow Mg_2P_2O_7 + 2NH_3 + H_2O$.

The precautions mentioned on pages 111 and 123 must be observed with great care during the ignition of this precipitate. The danger here lies in a possible reduction of the phosphate by the carbon of the filter paper, or by the ammonia evolved, which may act as a reducing agent. The phosphorus then attacks and injures a platinum crucible, and the determination is valueless.]

ANALYSIS OF LIMESTONE

Limestones vary widely in composition from a nearly pure marble through the dolomitic limestones, containing varying amounts of magnesium, to the impure varieties, which contain also ferrous and manganous carbonates and siliceous compounds in variable proportions. Many other minerals may be inclosed in limestones in small quantities, and an exact qualitative analysis will often show the presence of sulphides or sulphates, phosphates, and titanates, and the alkali or even the heavy metals. No attempt is made in the following procedures to provide a complete quantitative scheme which would take into account all of these constituents. Such a scheme for a complete analysis of a limestone may be found in Bulletin No. 700 of the United States Geological Survey. It is assumed that, for these practice determinations, a limestone is selected which contains only the more common constituents first enumerated above.

DETERMINATION OF MOISTURE

The determination of the amount of moisture in minerals or ores is often of great importance. Ores which have been exposed to the weather during shipment may have absorbed enough moisture to appreciably affect the results of analysis. Since it is essential that the seller and buyer should make their analyses upon comparable material, it is customary for each analyst to determine the moisture in the sample examined, and then to calculate the percentages of the various constituents with reference to a sample dried in the air, or at a temperature a little above 100°C., which, unless the ore has undergone chemical change because of the wetting, should be the same before and after shipment.

PROCEDURE.—Spread 25 grams of the powdered sample on a weighed watch-glass; weigh to the nearest 10 milligrams only and heat at 105°C.; weigh at intervals of an hour, after cooling in a desiccator, until the loss of weight after an hour's heating does not exceed 10 milligrams. It should be noted that a variation in weight of 10 milligrams in a total weight of 25 grams is no greater relatively than a variation of 0.1 milligram when the sample taken weighs 0.25 gram

DETERMINATION OF THE INSOLUBLE MATTER AND SILICA

PROCEDURE.—Weigh out two portions of the original powdered sample (not the dried sample), of about 5 grams each, into 250 cc. casseroles, and cover each with a watch-glass (Note 1). Pour over the powder 25 cc. of water, and then add 50 cc. of dilute hydrochloric acid (sp. gr. 1.12) in small portions, warming gently, until nothing further appears to dissolve (Note 2). Evaporate to dryness on the water bath. Pour over the residue a mixture of 5 cc. of water and 5 cc. of concentrated hydrochloric acid (sp. gr. 1.2) and again evaporate to dryness, and finally heat for at least an hour at a temperature of 110°C. Pour over this residue 50 cc. of dilute hydrochloric acid (one volume acid (sp. gr. 1.12) to five volumes water), and boil for about five minutes; then filter and wash twice with the dilute hydrochloric acid, and then with hot water until free from chlorides. Transfer the filter and contents to a porcelain crucible, dry carefully over a low flame, and ignite to constant weight. The residue represents the insoluble matter and the silica from any soluble silicates (Note 3).

Calculate the combined percentage of these in the limestone.

[Note 1: The relatively large weight (5 grams) taken for analysis insures greater accuracy in the determination of the ingredients which are present in small proportions, and is also more likely to be a representative sample of the material analyzed.]

[Note 2: It is plain that the amount of the insoluble residue and also its character will often depend upon the strength of acid used for solution of the limestone. It cannot, therefore, be regarded as representing any well-defined constituent, and its determination is essentially empirical.]

[Note 3: It is probable that some of the silicates present are wholly or partly decomposed by the acid, and the soluble silicic acid must be converted by evaporation to dryness, and heating, into white, insoluble silica. This change is not complete after one evaporation. The heating at a temperature somewhat higher than that of the water bath for a short time tends to leave the silica in the form of a powder, which promotes subsequent filtration. The siliceous residue is washed first with dilute acid to prevent hydrolytic changes, which would result in the formation of appreciable quantities of insoluble basic iron or aluminium salts on the filter when washing with hot water.

If it is desired to determine the percentage of silica separately, the ignited residue should be mixed in a platinum crucible with about six times its weight of anhydrous sodium carbonate, and the procedure given on page 151 should be followed. The filtrate from the silica is then added to the main filtrate from the insoluble residue.]

DETERMINATION OF FERRIC OXIDE AND ALUMINIUM OXIDE (WITH MANGANESE)

PROCEDURE. — To the filtrate from the insoluble residue add ammonium hydroxide until the solution just smells distinctly of ammonia, but do not add an excess. Then add 5 cc. of saturated bromine water (Note 1), and boil for five minutes. If the smell of ammonia has disappeared, again add ammonium hydroxide in slight excess, and 3 cc. of bromine water, and heat again for a few minutes. Finally add 10 cc. of ammonium chloride solution and keep the solution warm until it barely smells of ammonia; then filter promptly (Note 2). Wash the filter twice with hot water, then (after replacing the receiving beaker) pour through it 25 cc. of hot, dilute hydrochloric acid (one volume dilute HCl [sp. gr. 1.12] to five volumes water). A brown residue insoluble in the acid may be allowed to remain on the filter. Wash the filter five times with hot water, add to the filtrate ammonium hydroxide and bromine water as described above, and repeat the precipitation. Collect the precipitate on the filter already used, wash it free from chlorides with hot water, and ignite and weigh as described for ferric hydroxide on page 110. The residue after ignition consists of ferric oxide, alumina, and mangano-manganic oxide (Mn_3O_4), if manganese is present. These are commonly determined together (Note 3).

Calculate the percentage of the combined oxides in the limestone.

[Note 1: The addition of bromine water to the ammoniacal solutions serves to oxidize any ferrous hydroxide to ferric hydroxide and to precipitate manganese as $MnO(OH)_2$. The solution must contain not more than a bare excess of hydroxyl ions (ammonium hydroxide) when it is filtered, on account of the tendency of the aluminium hydroxide to redissolve.

The solution should not be strongly ammoniacal when the bromine is added, as strong ammonia reacts with the bromine, with the evolution of nitrogen.]

[Note 2: The precipitate produced by ammonium hydroxide and bromine should be filtered off promptly, since the alkaline solution absorbs carbon dioxide from the air, with consequent partial precipitation of the calcium as carbonate. This is possible even under the most favorable conditions, and for this reason the iron precipitate is redissolved and again precipitated to free it from calcium. When the precipitate is small, this reprecipitation may be omitted.]

[Note 3: In the absence of significant amounts of manganese the iron and aluminium may be separately determined by fusion of the mixed ignited precipitate, after weighing, with about ten times its weight of acid potassium sulphate, solution of the cold fused mass in water, and volumetric determination of the iron, as described on page 66. The aluminium is then determined by difference, after subtracting the weight of ferric oxide corresponding to the amount of iron found.

If a separate determination of the iron, aluminium, and manganese is desired, the mixed precipitate may be dissolved in acid before ignition, and the separation effected by special methods (see, for example, Fay, !Quantitative Analyses!, First Edition, pp. 15-19 and 23-27).]

DETERMINATION OF CALCIUM

PROCEDURE.—To the combined filtrates from the double precipitation of the hydroxides just described, add 5 cc. of dilute ammonium hydroxide (sp. gr. 0.96), and transfer the liquid to a 500 cc. graduated flask, washing out the beaker carefully. Cool to laboratory temperature, and fill the flask with distilled water until the lowest point of the meniscus is exactly level with the mark on the neck of the flask. Carefully remove any drops of water which are on the inside of the neck of the flask above the graduation by means of a strip of filter paper, make the solution uniform by pouring it out into a dry beaker and back into the flask several times. Measure off one fifth of this solution as follows (Note 1): Pour into a 100 cc. graduated flask about 10 cc. of the solution, shake the liquid thoroughly over the inner surface of the small flask, and pour it out. Repeat the same operation. Fill the 100 cc. flask until the lowest point of the meniscus is exactly level with the mark on its neck, remove any drops of solution from the upper part of the neck with filter paper, and pour the solution into a beaker (400-500 cc.). Wash out the flask with small quantities of water until it is clean, adding these to the 100 cc. of solution. When the duplicate portion of 100 cc. is measured out from the solution, remember that the flask must be rinsed out twice with that solution, as prescribed above, before the measurement is made. (A 100 cc. pipette may be used to measure out the aliquot portions, if preferred.)

Dilute each of the measured portions to 250 cc. with distilled water, heat the whole to boiling, and add ammonium oxalate solution slowly in moderate excess, stirring well. Boil for two minutes; allow the precipitated calcium oxalate to settle for a half-hour, and decant through a filter. Test the filtrate for complete precipitation by adding a few cubic centimeters of the precipitant, allowing it to stand for fifteen minutes. If no precipitate forms, make the solution slightly acid with hydrochloric acid (Note 2); see that it is properly labeled and reserve it to be combined with the filtrate from the second calcium oxalate precipitation (Notes 3 and 4).

Redissolve the calcium oxalate in the beaker with warm hydrochloric acid, pouring the acid through the filter. Wash the filter five times with water, and finally pour through it aqueous ammonia. Dilute the solution to 250 cc., bring to boiling, and add 1 cc. ammonium oxalate solution (Note 5) and ammonia in slight excess; boil for two minutes, and set aside for a half-hour. Filter off the calcium oxalate upon the filter first used, and wash free from chlorides. The filtrate should be made barely acid with hydrochloric acid and combined with the filtrate from the first precipitation. Begin at once the evaporation of the solutions for the determination of magnesium as described below.

The precipitate of calcium oxalate may be converted into calcium oxide by ignition without previous drying. After burning the filter, it may be ignited for three quarters of an hour in a platinum crucible at the highest heat of the Bunsen or Tirrill burner, and finally for ten minutes at the blast lamp (Note 6). Repeat the heating over the blast lamp until the weight is constant. As the calcium oxide absorbs moisture from the air, it must (after cooling) be weighed as rapidly as possible.

The precipitate may, if preferred, be placed in a weighted porcelain crucible. After burning off the filter and heating for ten minutes the calcium precipitate may be converted into calcium sulphate by placing 2 cc. of dilute sulphuric acid in the crucible (cold), heating the covered crucible very cautiously over a low flame to drive off the excess of acid, and finally at redness to constant weight (Note 7).

From the weight of the oxide or sulphate, calculate the percentage of the calcium (Ca) in the limestone, remembering that only one fifth of the total solution is used for this determination.

[Note 1: If the calcium were precipitated from the entire solution, the quantity of the precipitate would be greater than could be properly treated. The solution is, therefore, diluted to a definite volume (500 cc.), and exactly one fifth (100 cc.) is measured off in a graduated flask or by means of a pipette.]

[Note 2: The filtrate from the calcium oxalate should be made slightly acid immediately after filtration, in order to avoid the solvent action of the alkaline liquid upon the glass.]

[Note 3: The accurate quantitative separation of calcium and magnesium as oxalates requires considerable care. The calcium precipitate usually carries down with it some magnesium, and this can best be removed by redissolving the precipitate after filtration, and reprecipitation in the presence of only the small amount of magnesium which was included in the first precipitate. When, however, the proportion of magnesium is not very large, the second precipitation of the calcium can usually be avoided by precipitating it from a rather dilute solution (800 cc. or so) and in the presence of a considerable excess of the precipitant, that is, rather more than enough to convert both the magnesium and calcium into oxalates.]

[Note 4: The ionic changes involved in the precipitation of calcium as oxalate are exceedingly simple, and the principles discussed in connection with the barium sulphate precipitation on page 113 also apply here. The reaction is

$C_2O_4^{--} + Ca^{++} \rightarrow [CaC_2O_4].$

Calcium oxalate is nearly insoluble in water, and only very slightly soluble in acetic acid, but is readily dissolved by the strong mineral acids. This behavior with acids is explained by the fact that oxalic acid is a stronger acid than acetic acid; when, therefore, the oxalate is brought into contact with the latter there is almost no tendency to diminish the concentration of $C_2O_4^{--}$ ions by the formation of an acid less dissociated than the acetic acid itself, and practically no solvent action ensues. When a strong mineral acid is present, however, the ionization of the oxalic acid is much reduced by the high concentration of the H^{+} ions from the strong acid, the formation of the undissociated acid lessens the concentration of the $C_2O_4^{--}$ ions in solution, more of the oxalate passes into solution to re-establish equilibrium, and this process repeats itself until all is dissolved.

The oxalate is immediately reprecipitated from such a solution on the addition of OH^{-} ions, which, by uniting with the H^{+} ions of the acids (both the mineral acid and the oxalic acid) to form water, leave the Ca^{++} and $C_2O_4^{--}$ ions in the solution to recombine to form [CaC_2O_4], which is precipitated in the absence of the H^{+} ions. It is well at this point to add a small excess of $C_2O_4^{-}$ ions in the form of ammonium oxalate to decrease the solubility of the precipitate.

The oxalate precipitate consists mainly of $CaC_2O_4 \cdot H_2O$ when thrown down.]

[Note 5: The small quantity of ammonium oxalate solution is added before the second precipitation of the calcium oxalate to insure the presence of a slight excess of the reagent, which promotes the separation of the calcium compound.]

[Note 6: On ignition the calcium oxalate loses carbon dioxide and carbon monoxide, leaving calcium oxide:

$CaC_2O_4 \cdot H_2O \longrightarrow CaO + CO_2 + CO + H_2O$.

For small weights of the oxalate (0.6 gram or less) this reaction may be brought about in a platinum crucible at the highest temperature of a Tirrill burner, but it is well to ignite larger quantities than this over the blast lamp until the weight is constant.]

[Note 7: The heat required to burn the filter, and that subsequently applied as described, will convert most of the calcium oxalate to calcium carbonate, which is changed to sulphate by the sulphuric acid. The reactions involved are

$CaC_2O_4 \longrightarrow CaCO_3 + CO$,
$CaCO_3 + H_2SO_4 \longrightarrow CaSO_4 + H_2O + CO_2$.

If a porcelain crucible is employed for ignition, this conversion to sulphate is to be preferred, as a complete conversion to oxide is difficult to accomplish.]

[Note 8: The determination of the calcium may be completed volumetrically by washing the calcium oxalate precipitate from the filter into dilute sulphuric acid, warming, and titrating the liberated oxalic acid with a standard solution of potassium permanganate as described on page 72. When a considerable number of analyses are to be made, this procedure will save much of the time otherwise required for ignition and weighing.]

DETERMINATION OF MAGNESIUM

PROCEDURE. — Evaporate the acidified filtrates from the calcium precipitates until the salts begin to crystallize, but do !not! evaporate to dryness (Note 1). Dilute the solution cautiously until the salts are brought into solution, adding a little acid if the solution has evaporated to very small volume. The solution should be carefully examined at this point and must be filtered if a precipitate has appeared. Heat the clear solution to boiling; remove the burner and add 25 cc. of a solution of disodium phosphate. Then add slowly dilute ammonia (1 volume strong ammonia (sp. gr. 0.90) and 9 volumes water) as long as a precipitate continues to form. Finally, add a volume of concentrated ammonia (sp. gr. 0.90) equal to one third of the volume of the solution, and allow the whole to stand for about twelve hours.

Decant the solution through a filter, wash it with dilute ammonia water, proceeding as prescribed for the determination of phosphoric anhydride on page 122, including; the reprecipitation (Note 2), except that 3 cc. of disodium phosphate solution are added before the reprecipitation of the magnesium ammonium phosphate instead of the magnesia mixture there prescribed. From the weight of the pyrophosphate, calculate the percentage of magnesium oxide (MgO) in the sample of limestone. Remember that the pyrophosphate finally obtained is from one fifth of the original sample.

[Note 1: The precipitation of the magnesium should be made in as small volume as possible, and the ratio of ammonia to the total volume of solution should be carefully provided for, on account of the relative solubility of the magnesium ammonium phosphate. This matter has been fully discussed in connection with the phosphoric anhydride determination.]

[Note 2: The first magnesium ammonium phosphate precipitate is rarely wholly crystalline, as it should be, and is not always of the proper composition when precipitated in the presence of such large amounts of ammonium salts. The difficulty can best be remedied by filtering the precipitate and (without washing it) redissolving in a small quantity of hydrochloric acid, from which it may be again thrown down by ammonia after adding a little disodium phosphate solution. If the flocculent character was occasioned by the presence of magnesium hydroxide, the second precipitation, in a smaller volume containing fewer salts, will often result more favorably.

The removal of iron or alumina from a contaminated precipitate is a matter involving a long procedure, and a redetermination of the magnesium from a new sample, with additional precautions, is usually to be preferred.]

DETERMINATION OF CARBON DIOXIDE

!Absorption Apparatus!

[Illustration: Fig. 3]

The apparatus required for the determination of the carbon dioxide should be arranged as shown in the cut (Fig. 3). The flask (A) is an ordinary wash bottle, which should be nearly filled with dilute hydrochloric acid (100 cc. acid (sp. gr. 1.12) and 200 cc. of water). The flask is connected by rubber tubing (a) with the glass tube (b) leading nearly to the bottom of the evolution flask (B) and having its lower end bent upward and drawn out to small bore, so that the carbon dioxide evolved from the limestone cannot bubble back into (b). The evolution flask should preferably be a wide-mouthed Soxhlet extraction flask of about 150 cc. capacity because of the ease with which tubes and stoppers may be fitted into the neck of a flask of this type. The flask should be fitted with a two-hole rubber stopper. The condenser (C) may consist of a tube with two or three large bulbs blown in it, for use as an air-cooled condenser, or it may be a small water-jacketed condenser. The latter is to be preferred if a number of determinations are to be made in succession.

A glass delivery tube (c) leads from the condenser to the small U-tube (D) containing some glass beads or small pieces of glass rod and 3 cc. of a saturated solution of silver sulphate, with 3 cc. of concentrated sulphuric acid (sp. gr. 1.84). The short rubber tubing (d) connects the first U-tube to a second U-tube (E) which is filled with small dust-free lumps of dry calcium chloride, with a small, loose plug of cotton at the top of each arm. Both tubes should be closed by cork stoppers, the tops of which are cut off level with, or preferably forced a little below, the top of the U-tube, and then neatly sealed with sealing wax.

The carbon dioxide may be absorbed in a tube containing soda lime (F) or in a Geissler bulb (F') containing a concentrated solution of potassium hydroxide (Note 2). The tube (F) is a glass-stoppered side-arm U-tube in which the side toward the evolution flask and one half of the other side are filled with small, dust-free lumps of soda lime of good quality (Note 3). Since soda lime contains considerable moisture, the other half of the right side of the tube is filled with small lumps of dry, dust-free calcium chloride to retain the moisture from the soda lime. Loose plugs of cotton are placed at the top of each arm and between the soda lime and the calcium chloride.

The Geissler bulb (F'), if used, should be filled with potassium hydroxide solution (1 part of solid potassium hydroxide dissolved in two parts of water) until each small bulb is about two thirds full (Note 4). A small tube containing calcium chloride is connected with the Geissler bulb proper by a ground joint and should be wired to the bulb for safety. This is designed to retain any moisture from the hydroxide solution. A piece of clean, fine copper wire is so attached to the bulb that it can be hung from the hook above a balance pan, or other support.

The small bottle (G) with concentrated sulphuric acid (sp. gr. 1.84) is so arranged that the tube (f) barely dips below the surface. This will prevent the absorption of water vapor by (F) or (F') and serves as an aid in regulating the flow of air through the apparatus. (H) is an aspirator bottle of about four liters capacity, filled with water; (k) is a safety tube and a means of refilling (H); (h) is a screw clamp, and (K) a U-tube filled with soda lime.

[Note 1: The air current, which is subsequently drawn through the apparatus, to sweep all of the carbon dioxide into the absorption apparatus, is likely to carry with it some hydrochloric acid from the evolution flask. This acid is retained by the silver sulphate solution. The addition of concentrated sulphuric acid to this solution reduces its vapor pressure so far that very little water is carried on by the air current, and this slight amount is absorbed by the calcium chloride in (E). As the calcium chloride frequently contains a small amount of a basic material which would absorb carbon dioxide, it is necessary to pass carbon dioxide through (E) for a short time and then drive all the gas out with a dry air current for thirty minutes before use.]

[Note 2: Soda-lime absorption tubes are to be preferred if a satisfactory quality of soda lime is available and the number of determinations to be made successively is small. The potash bulbs will usually permit of a larger number of successive determinations without refilling, but they require greater care in handling and in the analytical procedure.]

[Note 3: Soda lime is a mixture of sodium and calcium hydroxides. Both combine with carbon dioxide to form carbonates, with the evolution of water. Considerable heat is generated by the reaction, and the temperature of the tube during absorption serves as a rough index of the progress of the reaction through the mass of soda lime.

It is essential that soda lime of good quality for analytical purposes should be used. The tube should not contain dust, as this is likely to be swept away.]

[Note 4: The solution of the hydroxide for use in the Geissler bulb must be highly concentrated to insure complete absorption of the carbon dioxide and also to reduce the vapor pressure of the solution, thus lessening the danger of loss of water with the air which passes through the bulbs. The small quantity of moisture which is then carried out of the bulbs is held by the calcium chloride in the prolong tube. The best form of absorption bulb is that to which the prolong tube is attached by a ground glass joint.

After the potassium hydroxide is approximately half consumed in the first bulb of the absorption apparatus, potassium bicarbonate is formed, and as it is much less soluble than the carbonate, it often precipitates. Its formation is a warning that the absorbing power of the hydroxide is much diminished.]

!The Analysis!

PROCEDURE.— Weigh out into the flask (B) about 1 gram of limestone. Cover it with 15 cc. of water. Weigh the absorption apparatus (F) or (F') accurately after allowing it to stand for 30 minutes in the balance case, and wiping it carefully with a lintless cloth, taking care to handle it as little as possible after wiping (Note 1). Connect the absorption apparatus with (e) and (f). If a soda-lime tube is used, be sure that the arm containing the soda lime is next the tube (E) and that the glass stopcocks are open.

To be sure that the whole apparatus is airtight, disconnect the rubber tube from the flask (A), making sure that the tubes (a) and (b) do not contain any hydrochloric acid, close the pinchcocks (a) and (k) and open (h). No bubbles should pass through (D) or (G) after a few seconds. When assured that the fittings are tight, close (h) and open (a) cautiously to admit air to restore atmospheric pressure. This precaution is essential, as a sudden inrush of air will project liquid from (D) or (F'). Reconnect the rubber tube with the flask (A). Open the pinchcocks (a) and (k) and blow over about 10 cc. of the hydrochloric acid from (A) into (B). When the action of the acid slackens, blow over (slowly) another 10 cc.

The rate of gas evolution should not exceed for more than a few seconds that at which about two bubbles per second pass through (G) (Note 2). Repeat the addition of acid in small portions until the action upon the limestone seems to be at an end, taking care to close (a) after each addition of acid (Note 3). Disconnect (A) and connect the rubber tubing with the soda-lime tube (K) and open (a). Then close (k) and open (h), regulating the flow of water from (H) in such a way that about two bubbles per second pass through (G). Place a small flame under (B) and !slowly! raise the contents to boiling and boil for three minutes. Then remove the burner from under (B) and continue to draw air through the apparatus for 20-30 minutes, or until (H) is emptied (Note 4). Remove the absorption apparatus, closing the stopcocks on (F) or stoppering the open ends of (F'), leave the apparatus in the balance case for at least thirty minutes, wipe it carefully and weigh, after opening the stopcocks (or removing plugs). The increase in weight is due to absorption of CO_2, from which its percentage in the sample may be calculated.

After cleaning (B) and refilling (H), the apparatus is ready for the duplicate analysis.

[Note 1: The absorption tubes or bulbs have large surfaces on which moisture may collect. By allowing them to remain in the balance case for some time before weighing, the amount of moisture absorbed on the surface is as nearly constant as practicable during two weighings, and a uniform temperature is also assured. The stopcocks of the U-tube should be opened, or the plugs used to close the openings of the Geissler bulb should be removed before weighing in order that the air contents shall always be at atmospheric pressure.]

[Note 2: If the gas passes too rapidly into the absorption apparatus, some carbon dioxide may be carried through, not being completely retained by the absorbents.]

[Note 3: The essential ionic changes involved in this procedure are the following: It is assumed that the limestone, which is typified by calcium carbonate, is very slightly soluble in water, and the ions resulting are Ca^{++} and CO_3^{-}. In the presence of H^{+} ions of the mineral acid, the CO_3^{-} ions form $[H_2CO_3]$. This is not only a weak acid which, by its formation, diminishes the concentration of the CO_3^{-} ions, thus causing more of the carbonate to dissolve to re-establish equilibrium, but it is also an unstable compound and breaks down into carbon dioxide and water.]

[Note 4: Carbon dioxide is dissolved by cold water, but the gas is expelled by boiling, and, together with that which is distributed through the apparatus, is swept out into the absorption bulb by the current of air. This air is purified by drawing it through the tube (K) containing soda lime, which removes any carbon dioxide which may be in it.]

DETERMINATION OF LEAD, COPPER, IRON, AND ZINC IN BRASS

ELECTROLYTIC SEPARATIONS

!General Discussion!

When a direct current of electricity passes from one electrode to another through solutions of electrolytes, the individual ions present in these solutions tend to move toward the electrode of opposite electrical charge to that which each ion bears, and to be discharged by that electrode. Whether or not such discharge actually occurs in the case of any particular ion depends upon the potential (voltage) of the current which is passing through the solution, since for each ion there is, under definite conditions, a minimum potential below which the discharge of the ion cannot be effected. By taking advantage of differences in discharge-potentials, it is possible to effect separations of a number of the metallic ions by electrolysis, and at the same time to deposit the metals in forms which admit of direct weighing. In this way the slower procedures of precipitation and filtration may frequently be avoided. The following paragraphs present a brief statement of the fundamental principles and conditions underlying electro-analysis.

The total energy of an electric current as it passes through a solution is distributed among three factors, first, its potential, which is measured in volts, and corresponds to what is called "head" in a stream of water; second, current strength, which is measured in amperes, and corresponds to the volume of water passing a cross-section of a stream in a given time interval; and third, the resistance of the conducting medium, which is measured in ohms. The relation between these three factors is expressed by Ohm's law, namely, that !$I = E/R$!, when I is current strength, E potential, and R resistance. It is plain that, for a constant resistance, the strength of the current and its potential are mutually and directly interdependent.

As already stated, the applied electrical potential determines whether or not deposition of a metal upon an electrode actually occurs. The current strength determines the rate of deposition and the physical characteristics of the deposit. The resistance of the solution is generally so small as to fall out of practical consideration.

Approximate deposition-potentials have been determined for a number of the metallic elements, and also for hydrogen and some of the acid-forming radicals. The values given below are those required for deposition from normal solutions at ordinary temperatures with reference to a hydrogen electrode. They must be regarded as approximate, since several disturbing factors and some secondary reactions render difficult their exact application under the conditions of analysis. They are:

Zn Cd Fe Ni Pb H Cu Sb Hg Ag SO_4 +0.77 +0.42 +0.34 +0.33 +0.13 0 -0.34 -0.67 -0.76 -0.79 +1.90

From these data it is evident that in order to deposit copper from a normal solution of copper sulphate a minimum potential equal to the algebraic sum of the deposition-potentials of copper ions and sulphate ions must be applied, that is, +1.56 volts. The deposition of zinc from a solution of zinc sulphate would require +2.67 volts, but, since the deposition of hydrogen from sulphuric acid solution requires only +1.90 volts, the quantitative deposition of zinc by electrolysis from a sulphuric acid solution of a zinc salt is not practicable. On the other hand, silver, if present in a solution of copper sulphate, would deposit with the copper.

The foregoing examples suffice to illustrate the application of the principle of deposition potentials, but it must further be noted that the values stated apply to normal solutions of the compounds in question, that is, to solutions of considerable concentrations. As the concentration of the ions diminishes, and hence fewer ions approach the electrodes, somewhat higher voltages are required to attract and discharge them. From this it follows that the concentrations should be kept as high as possible to effect complete deposition in the least practicable time, or else the potentials applied must be progressively increased as deposition proceeds. In practice, the desired result is obtained by starting with small volumes of solution, using as large an electrode surface as possible, and by stirring the solution to bring the ions in contact with the electrodes. This is, in general, a more convenient procedure than that of increasing the potential of the current during electrolysis, although that method is also used.

As already stated, those ions in a solution of electrolytes will first be discharged which have the lowest deposition potentials, and so long as these ions are present around the electrode in considerable concentration they, almost alone, are discharged, but, as their concentration diminishes, other ions whose deposition potentials are higher but still within that of the current applied, will also begin to separate. For example, from a nitric acid solution of copper nitrate, the copper ions will first be discharged at the cathode, but as they diminish in concentration hydrogen ions from the acid (or water) will be also discharged. Since the hydrogen thus liberated is a reducing agent, the nitric acid in the solution is slowly reduced to ammonia, and it may happen that if the current is passed through for a long time, such a solution will become alkaline. Oxygen is liberated at the anode, but, since there is no oxidizable substance present around that electrode, it escapes as oxygen gas. It should be noted that, in general, the changes occurring at the cathode are reductions, while those at the anode are oxidations.

For analytical purposes, solutions of nitrates or sulphates of the metals are preferable to those of the chlorides, since liberated chlorine attacks the electrodes. In some cases, as for example, that of silver, solution of salts forming complex ions, like that of the double cyanide of silver and potassium, yield better metallic deposits.

Most metals are deposited as such upon the cathode; a few, notably lead and manganese, separate in the form of dioxides upon the anode. It is evidently important that the deposited material should be so firmly adherent that it can be washed, dried, and weighed without loss in handling. To secure these conditions it is essential that the current density (that is, the amount of current per unit of area of the electrodes) shall not be too high. In prescribing analytical conditions it is customary to state the current strength in "normal densities" expressed in amperes per 100 sq. cm. of electrode surface, as, for example, "$N.D_{100}$ = 2 amps."

If deposition occurs too rapidly, the deposit is likely to be spongy or loosely adherent and falls off on subsequent treatment. This places a practical limit to the current density to be employed, for a given electrode surface. The cause of the unsatisfactory character of the deposit is apparently sometimes to be found in the coincident liberation of considerable hydrogen and sometimes in the failure of the rapidly deposited material to form a continuous adherent surface. The effect of rotating electrodes upon the character of the deposit is referred to below.

The negative ions of an electrolyte are attracted to the anode and are discharged on contact with it. Anions such as the chloride ion yield chlorine atoms, from which gaseous chlorine molecules are formed and escape. The radicals which compose such ions as NO_3^{-} or SO_4^{-} are not capable of independent existence after discharge, and break down into oxygen and N_2O_5 and SO_3 respectively. The oxygen escapes and the anhydrides, reacting with water, re-form nitric and sulphuric acids.

The law of Faraday expresses the relation between current strength and the quantities of the decomposition products which, under constant conditions, appear at the electrodes, namely, that a given quantity of electricity, acting for a given time, causes the separation of chemically equivalent quantities of the various elements or radicals. For example, since 107.94 grams of silver is equivalent to 1.008 grams of hydrogen, and that in turn to 8 grams of oxygen, or 31.78 grams of copper, the quantity of electricity which will cause the deposit of 107.94 grams of silver in a given time will also separate the weights just indicated of the other substances. Experiments show that a current of one ampere passing for one second, i.e., a coulomb of electricity, causes the deposition of 0.001118 gram of silver from a normal solution of a silver salt. The number of coulombs required to deposit 107.94 grams is 107.94/0.001118 or 96,550 and the same number of coulombs will also cause the separation of 1.008 grams of hydrogen, 8 grams of oxygen or 31.78 grams of copper. While it might at first appear that Faraday's law could thus be used as a basis for the calculation of the time required for the deposition of a given quantity of an electrolyte from solution, it must be remembered that the law expresses what occurs when the concentration of the ions in the solution is kept constant, as, for example, when the anode in a silver salt solution is a plate of metallic silver. Under the conditions of electro-analysis the concentration of the ions is constantly diminishing as deposition proceeds and the time actually required for complete deposition of a given weight of material by a current of constant strength is, therefore, greater than that calculated on the basis of the law as stated above.

The electrodes employed in electro-analysis are almost exclusively of platinum, since that metal alone satisfactorily resists chemical action of the electrolytes, and can be dried and weighed without change in composition. The platinum electrodes may be used in the form of dishes, foil or gauze. The last, on account of the ease of circulation of the electrolyte, its relatively large surface in proportion to its weight and the readiness with which it can be washed and dried, is generally preferred.

Many devices have been described by the use of which the electrode upon which deposition occurs can be mechanically rotated. This has an effect parallel to that of greatly increasing the electrode surface and also provides a most efficient means of stirring the solution. With such an apparatus the amperage may be increased to 5 or even 10 amperes and depositions completed with great rapidity and accuracy. It is desirable, whenever practicable, to provide a rotating or stirring device, since, for example, the time consumed in the deposition of the amount of copper usually found in analysis may be reduced from the 20 to 24 hours required with stationary electrodes, and unstirred solutions, to about one half hour.

DETERMINATION OF COPPER AND LEAD

PROCEDURE.—Weigh out two portions of about 0.5 gram each (Note 1) into tall, slender lipless beakers of about 100 cc. capacity. Dissolve the metal in a solution of 5 cc. of dilute nitric acid (sp. gr. 1.20) and 5 cc. of water, heating gently, and keeping the beaker covered. When the sample has all dissolved (Note 2), wash down the sides of the beaker and the bottom of the watch-glass with water and dilute the solution to about 50 cc. Carefully heat to boiling and boil for a minute or two to expel nitrous fumes.

Meanwhile, four platinum electrodes, two anodes and two cathodes, should be cleaned by dipping in dilute nitric acid, washing with water and finally with 95 per cent alcohol (Note 3). The alcohol may be ignited and burned off. The electrodes are then cooled in a desiccator and weighed. Connect the electrodes with the binding posts (or other device for connection with the electric circuit) in such a way that the copper will be deposited upon the electrode with the larger surface, which is made the cathode. The beaker containing the solution should then be raised into place from below the electrodes until the latter reach nearly to the bottom of the beaker. The support for the beaker must be so arranged that it can be easily raised or lowered.

If the electrolytic apparatus is provided with a mechanism for the rotation of the electrode or stirring of the electrolyte, proceed as follows: Arrange the resistance in the circuit to provide a direct current of about one ampere. Pass this current through the solution to be electrolyzed, and start the rotating mechanism. Keep the beaker covered as completely as possible, using a split watch-glass (or other device) to avoid loss by spattering. When the solution is colorless, which is usually the case after about 35 minutes, rinse off the cover glass, wash down the sides of the beaker, add about 0.30 gram of urea and continue the electrolysis for another five minutes (Notes 4 and 5).

If stationary electrodes are employed, the current strength should be about 0.1 ampere, which may, after 12 to 15 hours, be increased to 0.2 ampere. The time required for complete deposition is usually from 20 to 24 hours. It is advisable to add 5 cc. of nitric acid (sp. gr. 1.2) if the electrolysis extends over this length of time. No urea is added in this case.

When the deposition of the copper appears to be complete, stop the rotating mechanism and slowly lower the beaker with the left hand, directing at the same time a stream of water from a wash bottle on both electrodes. Remove the beaker, shut off the current, and, if necessary, complete the washing of the electrodes (Note 6). Rinse the electrodes cautiously with alcohol and heat them in a hot closet until the alcohol has just evaporated, but no longer, since the copper is likely to oxidize at the higher temperature. (The alcohol may be removed by ignition if care is taken to keep the electrodes in motion in the air so that the copper deposit is not too strongly heated at any one point.)

Test the solution in the beaker for copper as follows, remembering that it is to be used for subsequent determinations of iron and zinc: Remove about 5 cc. and add a slight excess of ammonia. Compare the mixture with some distilled water, holding both above a white surface. The solution should not show any tinge of blue. If the presence of copper is indicated, add the test portion to the main solution, evaporate the whole to a volume of about 100 cc., and again electrolyze with clean electrodes (Note 7).

After cooling the electrodes in a desiccator, weigh them and from the weight of copper on the cathode and of lead dioxide (PbO_2) on the anode, calculate the percentage of copper (Cu) and of lead (Pb) in the brass.

[Note 1: It is obvious that the brass taken for analysis should be untarnished, which can be easily assured, when wire is used, by scouring with emery. If chips or borings are used, they should be well mixed, and the sample for analysis taken from different parts of the mixture.]

[Note 2: If a white residue remains upon treatment of the alloy with nitric acid, it indicates the presence of tin. The material is not, therefore, a true brass. This may be treated as follows: Evaporate the solution to dryness, moisten the residue with 5 cc. of dilute nitric acid (sp. gr. 1.2) and add 50 cc. of hot water. Filter off the meta-stannic acid, wash, ignite in porcelain and weigh as SnO_2. This oxide is never wholly free from copper and must be purified for an exact determination. If it does not exceed 2 per cent of the alloy, the quantity of copper which it contains may usually be neglected.]

[Note 3: The electrodes should be freed from all greasy matter before using, and those portions upon which the metal will deposit should not be touched with the fingers after cleaning.]

[Note 4: Of the ions in solution, the H^+, Cu^{++}, Zn^{++}, and Fe^{+++} ions tend to move toward the cathode. The NO_3^- ions and the lead, probably in the form of PbO_2^{--} ions, move toward the anode. At the cathode the Cu^{++} ions are discharged and plate out as metallic copper. This alone occurs while the solution is relatively concentrated. Later on, H^+ ions are also discharged. In the presence of considerable quantities of H^+ ions, as in this acid solution, no Zn^{++} or Fe^{+++} ions are discharged because of their greater deposition potentials. At the anode the lead is deposited as PbO_2 and oxygen is evolved.

For the reasons stated on page 141 care must be taken that the solution does not become alkaline if the electrolysis is long continued.]

[Note 5: Urea reacts with nitrous acid, which may be formed in the solution as a result of the reducing action of the liberated hydrogen. Its removal promotes the complete precipitation of the copper. The reaction is

$CO(NH_2)_2 + 2HNO_2 \rightarrow CO_2 + 2N_2 + 3H_2O$.]

[Note 6: The electrodes must be washed nearly or quite free from the nitric acid solution before the circuit is broken to prevent re-solution of the copper.

If several solutions are connected in the same circuit it is obvious that some device must be used to close the circuit as soon as the beaker is removed.]

[Note 7: The electrodes upon which the copper has been deposited may be cleaned by immersion in warm nitric acid. To remove the lead dioxide, add a few crystals of oxalic acid to the nitric acid.]

DETERMINATION OF IRON

Most brasses contain small percentages of iron (usually not over 0.1 per cent) which, unless removed, is precipitated as phosphate and weighed with the zinc.

PROCEDURE. — To the solution from the precipitation of copper and lead by electrolysis, add dilute ammonia (sp. gr. 0.96) until the precipitate of zinc hydroxide which first forms re-dissolves, leaving only a slight red precipitate of ferric hydroxide. Filter off the iron precipitate, using a washed filter, and wash five times with hot water. Test a portion of the last washing with a dilute solution of ammonium sulphide to assure complete removal of the zinc.

The precipitate may then be ignited and weighed as ferric oxide, as described on page 110.

Calculate the percentage of iron (Fe) in the brass.

DETERMINATION OF ZINC

PROCEDURE.—Acidify the filtrate from the iron determination with dilute nitric acid. Concentrate it to 150 cc. Add to the cold solution dilute ammonia (sp. gr. 0.96) cautiously until it barely smells of ammonia; then add !one drop! of a dilute solution of litmus (Note 1), and drop in, with the aid of a dropper, dilute nitric acid until the blue of the litmus just changes to red. It is important that this point should not be overstepped. Heat the solution nearly to boiling and pour into it slowly a filtered solution of di-ammonium hydrogen phosphate[1] containing a weight of the phosphate about equal to twelve times that of the zinc to be precipitated. (For this calculation the approximate percentage of zinc is that found by subtracting the sum of the percentages of the copper, lead and iron from 100 per cent.) Keep the solution just below boiling for fifteen minutes, stirring frequently (Note 2). If at the end of this time the amorphous precipitate has become crystalline, allow the solution to cool for about four hours, although a longer time does no harm (Note 3), and filter upon an asbestos filter in a porcelain Gooch crucible. The filter is prepared as described on page 103, and should be dried to constant weight at 105°C.

[Footnote 1: The ammonium phosphate which is commonly obtainable contains some mono-ammonium salt, and this is not satisfactory as a precipitant. It is advisable, therefore, to weigh out the amount of the salt required, dissolve it in a small volume of water, add a drop of phenolphthalein solution, and finally add dilute ammonium hydroxide solution cautiously until the solution just becomes pink, but do not add an excess.]

Wash the precipitate until free from sulphates with a warm 1 per cent solution of the di-ammonium phosphate, and then five times with 50 per cent alcohol (Note 4). Dry the crucible and precipitate for an hour at 105°C., and finally to constant weight (Note 5). The filtrate should be made alkaline with ammonia and tested for zinc with a few drops of ammonium sulphide, allowing it to stand (Notes 6, 7 and 8).

From the weight of the zinc ammonium phosphate ($ZnNH_4PO_4$) calculate the percentage of the zinc (Zn) in the brass.

[Note 1: The zinc ammonium phosphate is soluble both in acids and in ammonia. It is, therefore, necessary to precipitate the zinc in a nearly neutral solution, which is more accurately obtained by adding a drop of a litmus solution to the liquid than by the use of litmus paper.]

[Note 2: The precipitate which first forms is amorphous, and may have a variable composition. On standing it becomes crystalline and then has the composition $ZnNH_4PO_4$. The precipitate then settles rapidly and is apt to occasion "bumping" if the solution is heated to boiling. Stirring promotes the crystallization.]

[Note 3: In a carefully neutralized solution containing a considerable excess of the precipitant, and also ammonium salts, the separation of the zinc is complete after standing four hours. The ionic changes connected with the precipitation of the zinc as zinc ammonium phosphate are similar to those described for magnesium ammonium phosphate, except that the zinc precipitate is soluble in an excess of ammonium hydroxide, probably as a result of the formation of complex ions of the general character $Zn(NH_3)_4^{++}$.]

[Note 4: The precipitate is washed first with a dilute solution of the phosphate to prevent a slight decomposition of the precipitate (as a result of hydrolysis) if hot water alone is used. The alcohol is added to the final wash-water to promote the subsequent drying.]

[Note 5: The precipitate may be ignited and weighed as $Zn_2P_2O_7$, by cautiously heating the porcelain Gooch crucible within a nickel or iron crucible, used as a radiator. The heating must be very slow at first, as the escaping ammonia may reduce the precipitate if it is heated too quickly.]

[Note 6: If the ammonium sulphide produced a distinct precipitate, this should be collected on a small filter, dissolved in a few cubic centimeters of dilute nitric acid, and the zinc reprecipitated as phosphate, filtered off, dried, and weighed, and the weight added to that of the main precipitate.]

[Note 7: It has been found that some samples of asbestos are acted upon by the phosphate solution and lose weight. An error from this source may be avoided by determining the weight of the crucible and filter after weighing the precipitate. For this purpose the precipitate may be dissolved in dilute nitric acid, the asbestos washed thoroughly, and the crucible reweighed.]

[Note 8. The details of this method of precipitation of zinc are fully discussed in an article by Dakin, !Ztschr. Anal. Chem.!, 39 (1900), 273.]

DETERMINATION OF SILICA IN SILICATES

Of the natural silicates, or artificial silicates such as slags and some of the cements, a comparatively few can be completely decomposed by treatment with acids, but by far the larger number require fusion with an alkaline flux to effect decomposition and solution for analysis. The procedure given below applies to silicates undecomposable by acids, of which the mineral feldspar is taken as a typical example. Modifications of the procedure, which are applicable to silicates which are completely or partially decomposable by acids, are given in the Notes on page 155.

PREPARATION OF THE SAMPLE

Grind about 3 grams of the mineral in an agate mortar (Note 1) until no grittiness is to be detected, or, better, until it will entirely pass through a sieve made of fine silk bolting cloth. The sieve may be made by placing a piece of the bolting cloth over the top of a small beaker in which the ground mineral is placed, holding the cloth in place by means of a rubber band below the lip of the beaker. By inverting the beaker over clean paper and gently tapping it, the fine particles pass through the sieve, leaving the coarser particles within the beaker. These must be returned to the mortar and ground, and the process of sifting and grinding repeated until the entire sample passes through the sieve.

[Note 1: If the sample of feldspar for analysis is in the massive or crystalline form, it should be crushed in an iron mortar until the pieces are about half the size of a pea, and then transferred to a steel mortar, in which they are reduced to a coarse powder. A wooden mallet should always be used to strike the pestle of the steel mortar, and the blows should not be sharp.

It is plain that final grinding in an agate mortar must be continued until the whole of the portion of the mineral originally taken has been ground so that it will pass the bolting cloth, otherwise the sifted portion does not represent an average sample, the softer ingredients, if foreign matter is present, being first reduced to powder. For this reason it is best to start with not more than the quantity of the feldspar needed for analysis. The mineral must be thoroughly mixed after the grinding.]

FUSION AND SOLUTION

PROCEDURE.—Weigh into platinum crucibles two portions of the ground feldspar of about 0.8 gram each. Weigh on rough balances two portions of anhydrous sodium carbonate, each amounting to about six times the weight of the feldspar taken for analysis (Note 1). Pour about three fourths of the sodium carbonate into the crucible, place the latter on a piece of clean, glazed paper, and thoroughly mix the substance and the flux by carefully stirring for several minutes with a dry glass rod, the end of which has been recently heated and rounded in a flame and slowly cooled. The rod may be wiped off with a small fragment of filter paper, which may be placed in the crucible. Place the remaining fourth of the carbonate on the top of the mixture. Cover the crucible, heat it to dull redness for five minutes, and then gradually increase the heat to the full capacity of a Bunsen or Tirrill burner for twenty minutes, or until a quiet, liquid fusion is obtained (Note 2). Finally, heat the sides and cover strongly until any material which may have collected upon them is also brought to fusion.

Allow the crucible to cool, and remove the fused mass as directed on page 116. Disintegrate the mass by placing it in a previously prepared mixture of 100 cc. of water and 50 cc. of dilute hydrochloric acid (sp. gr. 1.12) in a covered casserole (Note 3). Clean the crucible and lid by means of a little hydrochloric acid, adding this acid to the main solution (Notes 4 and 5).

[Note 1: Quartz, and minerals containing very high percentages of silica, may require eight or ten parts by weight of the flux to insure a satisfactory decomposition.]

[Note 2: During the fusion the feldspar, which, when pure, is a silicate of aluminium and either sodium or potassium, but usually contains some iron, calcium, and magnesium, is decomposed by the alkaline flux. The sodium of the latter combines with the silicic acid of the silicate, with the evolution of carbon dioxide, while about two thirds of the aluminium forms sodium aluminate and the remainder is converted into basic carbonate, or the oxide. The calcium and magnesium, if present, are changed to carbonates or oxides.

The heat is applied gently to prevent a too violent reaction when fusion first takes place.]

[Note 3: The solution of a silicate by a strong acid is the result of the combination of the H^{+} ions of the acid and the silicate ions of the silicate to form a slightly ionized silicic acid. As a consequence, the concentration of the silicate ions in the solution is reduced nearly to zero, and more silicate dissolves to re-establish the disturbed equilibrium. This process repeats itself until all of the silicate is brought into solution.

Whether the resulting solution of the silicate contains ortho-silicic acid (H_4SiO_4) or whether it is a colloidal solution of some other less hydrated acid, such as meta-silicic acid (H_2SiO_3), is a matter that is still debatable. It is certain, however, that the gelatinous material which readily separates from such solutions is of the nature of a hydrogel, that is, a colloid which is insoluble in water. This substance when heated to 100°C., or higher, is completely dehydrated, leaving only the anhydride, SiO_2. The changes may be represented by the equation:

$$SiO_3^{-} + 2H^{+} \longrightarrow [H_2SiO_3] \longrightarrow H_2O + SiO_2.]$$

[Note 4: A portion of the fused mass is usually projected upward by the escaping carbon dioxide during the fusion. The crucible must therefore be kept covered as much as possible and the lid carefully cleaned.]

[Note 5: A gritty residue remaining after the disintegration of the fused mass by acid indicates that the substance has been but imperfectly decomposed. Such a residue should be filtered, washed, dried, ignited, and again fused with the alkaline flux; or, if the quantity of material at hand will permit, it is better to reject the analysis, and to use increased care in grinding the mineral and in mixing it with the flux.]

DEHYDRATION AND FILTRATION

PROCEDURE.—Evaporate the solution of the fusion to dryness, stirring frequently until the residue is a dry powder. Moisten the residue with 5 cc. of strong hydrochloric acid (sp. gr. 1.20) and evaporate again to dryness. Heat the residue for at least one hour at a temperature of 110°C. (Note 1). Again moisten the residue with concentrated hydrochloric acid, warm gently, making sure that the acid comes into contact with the whole of the residue, dilute to about 200 cc. and bring to boiling. Filter off the silica without much delay (Note 2), and wash five times with warm dilute hydrochloric acid (one part dilute acid (1.12 sp. gr.) to three parts of water). Allow the filter to drain for a few moments, then place a clean beaker below the funnel and wash with water until free from chlorides, discarding these washings. Evaporate the original filtrate to dryness, dehydrate at 110°C. for one hour (Note 3), and proceed as before, using a second filter to collect the silica after the second dehydration. Wash this filter with warm, dilute hydrochloric acid (Note 4), and finally with hot water until free from chlorides.

[Note 1: The silicic acid must be freed from its combination with a base (sodium, in this instance) before it can be dehydrated. The excess of hydrochloric acid accomplishes this liberation. By disintegrating the fused mass with a considerable volume of dilute acid the silicic acid is at first held in solution to a large extent. Immediate treatment of the fused mass with strong acid is likely to cause a semi-gelatinous silicic acid to separate at once and to inclose alkali salts or alumina.

A flocculent residue will often remain after the decomposition of the fused mass is effected. This is usually partially dehydrated silicic acid and does not require further treatment at this point. The progress of the dehydration is indicated by the behavior of the solution, which as evaporation proceeds usually gelatinizes. On this account it is necessary to allow the solution to evaporate on a steam bath, or to stir it vigorously, to avoid loss by spattering.]

[Note 2: To obtain an approximately pure silica, the residue after evaporation must be thoroughly extracted by warming with hydrochloric acid, and the solution freely diluted to prevent, as far as possible, the inclosure of the residual salts in the particles of silica.

The filtration should take place without delay, as the dehydrated silica slowly dissolves in hydrochloric acid on standing.]

[Note 3: It has been shown by Hillebrand that silicic acid cannot be completely dehydrated by a single evaporation and heating, nor by several such treatments, unless an intermediate filtration of the silica occurs. If, however, the silica is removed and the filtrates are again evaporated and the residue heated, the amount of silica remaining in solution is usually negligible, although several evaporations and filtrations are required with some silicates to insure absolute accuracy.

It is probable that temperatures above 100°C. are not absolutely necessary to dehydrate the silica; but it is recommended, as tending to leave the silica in a better condition for filtration than when the lower temperature of the water bath is used. This, and many other points in the analysis of silicates, are fully discussed by Dr. Hillebrand in the admirable monograph on "The Analysis of Silicate and Carbonate Rocks," Bulletin No. 700 of the United States Geological Survey.

The double evaporation and filtration spoken of above are essential because of the relatively large amount of alkali salts (sodium chloride) present after evaporation. For the highest accuracy in the determination of silica, or of iron and alumina, it is also necessary to examine for silica the precipitate produced in the filtrate by ammonium hydroxide by fusing it with acid potassium sulphate and solution of the fused mass in water. The insoluble silica is filtered, washed, and weighed, and the weight added to the weight of silica previously obtained.]

[Note 4: Aluminium and iron are likely to be thrown down as basic salts from hot, very dilute solutions of their chlorides, as a result of hydrolysis. If the silica were washed only with hot water, the solution of these chlorides remaining in the filter after the passage of the original filtrate would gradually become so dilute as to throw down basic salts within the pores of the filter, which would remain with the silica. To avoid this, an acid wash-water is used until the aluminium and iron are practically removed. The acid is then removed by water.]

IGNITION AND TESTING OF SILICA

PROCEDURE.—Transfer the two washed filters belonging to each determination to a platinum crucible, which need not be previously weighed, and burn off the filter (Note 1). Ignite for thirty minutes over the blast lamp with the cover on the crucible, and then for periods of ten minutes, until the weight is constant.

When a constant weight has been obtained, pour into the crucible about 3 cc. of water, and then 3 cc. of hydrofluoric acid. !This must be done in a hood with a good draft and great care must be taken not to come into contact with the acid or to inhale its fumes (Note 2!).

If the precipitate has dissolved in this quantity of acid, add two drops of concentrated sulphuric acid, and heat very slowly (always under the hood) until all the liquid has evaporated, finally igniting to redness. Cool in a desiccator, and weigh the crucible and residue. Deduct this weight from the previous weight of crucible and impure silica, and from the difference calculate the percentage of silica in the sample (Note 3).

[Note 1: The silica undergoes no change during the ignition beyond the removal of all traces of water; but Hillebrand (!loc. cit.!) has shown that the silica holds moisture so tenaciously that prolonged ignition over the blast lamp is necessary to remove it entirely. This finely divided, ignited silica tends to absorb moisture, and should be weighed quickly.]

[Note 2: Notwithstanding all precautions, the ignited precipitate of silica is rarely wholly pure. It is tested by volatilisation of the silica as silicon fluoride after solution in hydrofluoric acid, and, if the analysis has been properly conducted, the residue, after treatment with the acids and ignition, should not exceed 1 mg.

The acid produces ulceration if brought into contact with the skin, and its fumes are excessively harmful if inhaled.]

[Note 3: The impurities are probably weighed with the original precipitate in the form of oxides. The addition of the sulphuric acid displaces the hydrofluoric acid, and it may be assumed that the resulting sulphates (usually of iron or aluminium) are converted to oxides by the final ignition.

It is obvious that unless the sulphuric and hydrofluoric acids used are known to leave no residue on evaporation, a quantity equal to that employed in the analysis must be evaporated and a correction applied for any residue found.]

[Note 4: If the silicate to be analyzed is shown by a previous qualitative examination to be completely decomposable, it may be directly treated with hydrochloric acid, the solution evaporated to dryness, and the silica dehydrated and further treated as described in the case of the feldspar after fusion.

A silicate which gelatinizes on treatment with acids should be mixed first with a little water, and the strong acid added in small portions with stirring, otherwise the gelatinous silicic acid incloses particles of the original silicate and prevents decomposition. The water, by separating the particles and slightly lessening the rapidity of action, prevents this difficulty. This procedure is one which applies in general to the solution of fine mineral powders in acids.

If a small residue remains undecomposed by the treatment of the silicate with acid, this may be filtered, washed, ignited and fused with sodium carbonate and a solution of the fused mass added to the original acid solution. This double procedure has an advantage, in that it avoids adding so large a quantity of sodium salts as is required for disintegration of the whole of the silicate by the fusion method.]

PART IV

STOICHIOMETRY

The problems with which the analytical chemist has to deal are not, as a matter of actual fact, difficult either to solve or to understand. That they appear difficult to many students is due to the fact that, instead of understanding the principles which underlie each of the small number of types into which these problems may be grouped, each problem is approached as an individual puzzle, unrelated to others already solved or explained. This attitude of mind should be carefully avoided.

It is obvious that ability to make the calculations necessary for the interpretation of analytical data is no less important than the manipulative skill required to obtain them, and that a moderate time spent in the careful study of the solutions of the typical problems which follow may save much later embarrassment.

1. It is often necessary to calculate what is known as a "chemical factor," or its equivalent logarithmic value called a "log factor," for the conversion of the weight of a given chemical substance into an equivalent weight of another substance. This is, in reality, a very simple problem in proportion, making use of the atomic or molecular weights of the substances in question which are chemically equivalent to each other. One of the simplest cases of this sort is the following: What is the factor for the conversion of a given weight of barium sulphate ($BaSO_4$) into an equivalent weight of sulphur (S)? The molecular weight of $BaSO_4$ is 233.5. There is one atom of S in the molecule and the atomic weight of S is 32.1. The chemical factor is, therefore, 32.1/233.5, or 0.1375 and the weight of S corresponding to a given weight of $BaSO_4$ is found by multiplying the weight of $BaSO_4$ by this factor. If the problem takes the form, "What is the factor for the conversion of a given weight of ferric oxide (Fe_2O_3) into ferrous oxide (FeO), or of a given weight of mangano-manganic oxide (Mn_3O_4) into manganese (Mn)?" the

principle involved is the same, but it must then be noted that, in the first instance, each molecule of Fe_2O_3 will be equivalent to two molecules of FeO, and in the second instance that each molecule of Mn_3O_4 is equivalent to three atoms of Mn. The respective factors then become

$(2FeO/Fe_2O_3)$ or $(143.6/159.6)$ and $(3Mn/Mn_3O_4)$ or $(164.7/228.7)$.

It is obvious that the arithmetical processes involved in this type of problem are extremely simple. It is only necessary to observe carefully the chemical equivalents. It is plainly incorrect to express the ratio of ferrous to ferric oxide as (FeO/Fe_2O_3), since each molecule of the ferric oxide will yield two molecules of the ferrous oxide. Mistakes of this sort are easily made and constitute one of the most frequent sources of error.

2. A type of problem which is slightly more complicated in appearance, but exactly comparable in principle, is the following: "What is the factor for the conversion of a given weight of ferrous sulphate ($FeSO_4$), used as a reducing agent against potassium permanganate, into the equivalent weight of sodium oxalate ($Na_2C_2O_4$)?" To determine the chemical equivalents in such an instance it is necessary to inspect the chemical reactions involved. These are:

$10FeSO_4 + 2KMnO_4 + 8H_2SO_4 \rightarrow 5Fe_2(SO_4)_3 + K_2SO_4 + 2MnSO_4 + 8H_2O$,

$5Na_2C_2O_4 + 2KMnO_4 + 8H_2SO_4 \rightarrow 5Na_2SO_4 + 10CO_2 + K_2SO_4 + 2MnSO_4 + 8H_2O$.

It is evident that $10FeSO_4$ in the one case, and $5Na_2C_2O_4$ in the other, each react with $2KMnO_4$. These molecular quantities are therefore equivalent, and the factor becomes $(10FeSO_4/5Na_2C_2O_4)$ or $(2FeSO_4/Na_2C_2O_4)$ or $(303.8/134)$.

Again, let it be assumed that it is desired to determine the factor required for the conversion of a given weight of potassium permanganate ($KMnO_4$) into an equivalent weight of potassium bichromate ($K_2Cr_2O_7$), each acting as an oxidizing agent against ferrous sulphate. The reactions involved are:

10FeSO$_4$ + 2KMnO$_4$ + 8H$_2$SO$_4$ —> 5Fe$_2$(SO$_4$)$_3$ + K$_2$SO$_4$ + 2MnSO$_4$ + 8H$_2$O,

6FeSO$_4$ + K$_2$Cr$_2$O$_7$ + 7H$_2$SO$_4$ —> 3Fe$_2$(SO$_3$)$_3$ + K$_2$SO$_4$ + Cr$_2$(SO$_4$)$_3$ + 7H$_2$O.

An inspection of these equations shows that 2KMO$_4$ react with 10FeSO$_4$, while K$_2$Cr$_2$O$_7$ reacts with 6FeSO$_4$. These are not equivalent, but if the first equation is multiplied by 3 and the second by 5 the number of molecules of FeSO$_4$ is then the same in both, and the number of molecules of KMnO$_4$ and K$_2$Cr$_2$O$_7$ reacting with these 30 molecules become 6 and 5 respectively. These are obviously chemically equivalent and the desired factor is expressed by the fraction (6KMnO$_4$/5K$_2$Cr$_2$O$_7$) or (948.0/1471.0).

3. It is sometimes necessary to calculate the value of solutions according to the principles just explained, when several successive reactions are involved. Such problems may be solved by a series of proportions, but it is usually possible to eliminate the common factors and solve but a single one. For example, the amount of MnO$_2$ in a sample of the mineral pyrolusite may be determined by dissolving the mineral in hydrochloric acid, absorbing the evolved chlorine in a solution of potassium iodide, and measuring the liberated iodine by titration with a standard solution of sodium thiosulphate. The reactions involved are:

MnO$_2$ + 4HCl —> MnCl$_2$ + 2H$_2$O + Cl$_2$
Cl$_2$ + 2KI —> I$_2$ + 2KCl
I$_2$ + 2Na$_2$S$_2$O$_3$ —> 2NaI + Na$_2$S$_4$O$_6$

Assuming that the weight of thiosulphate corresponding to the volume of sodium thiosulphate solution used is known, what is the corresponding weight of manganese dioxide? From the reactions given above, the following proportions may be stated:

2Na$_2$S$_2$O$_3$:I$_2$ = 316.4:253.9,

I$_2$:Cl$_2$ = 253.9:71,

Cl$_2$:MnO$_2$ = 71:86.9.

After canceling the common factors, there remains $2Na_2S_2O_3 : MnO_2 = 316.4 : 86.9$, and the factor for the conversion of thiosulphate into an equivalent of manganese dioxide is $86.9/316.4$.

4. To calculate the volume of a reagent required for a specific operation, it is necessary to know the exact reaction which is to be brought about, and, as with the calculation of factors, to keep in mind the molecular relations between the reagent and the substance reacted upon. For example, to estimate the weight of barium chloride necessary to precipitate the sulphur from 0.1 gram of pure pyrite (FeS_2), the proportion should read

488. $120.0 \; 2(BaCl_2 \cdot 2H_2O) : FeS_2 = x : 0.1$,

where !x! represents the weight of the chloride required. Each of the two atoms of sulphur will form upon oxidation a molecule of sulphuric acid or a sulphate, which, in turn, will require a molecule of the barium chloride for precipitation. To determine the quantity of the barium chloride required, it is necessary to include in its molecular weight the water of crystallization, since this is inseparable from the chloride when it is weighed. This applies equally to other similar instances.

If the strength of an acid is expressed in percentage by weight, due regard must be paid to its specific gravity. For example, hydrochloric acid (sp. gr. 1.12) contains 23.8 per cent HCl !by weight!; that is, 0.2666 gram HCl in each cubic centimeter.

5. It is sometimes desirable to avoid the manipulation required for the separation of the constituents of a mixture of substances by making what is called an "indirect analysis." For example, in the analysis of silicate rocks, the sodium and potassium present may be obtained in the form of their chlorides and weighed together. If the weight of such a mixture is known, and also the percentage of chlorine present, it is possible to calculate the amount of each chloride in the mixture. Let it be assumed that the weight of the mixed chlorides is 0.15 gram, and that it contains 53 per cent of chlorine.

The simplest solution of such a problem is reached through algebraic methods. The weight of chlorine is evidently 0.15 x 0.53, or

0.0795 gram. Let x represent the weight of sodium chloride present and y that of potassium chloride. The molecular weight of NaCl is 58.5 and that of KCl is 74.6. The atomic weight of chlorine is 35.5. Then

x + y = 0.15 (35.5/58.5)x + (35.5/74.6)y = 0.00795

Solving these equations for x shows the weight of NaCl to be 0.0625 gram. The weight of KCl is found by subtracting this from 0.15.

The above is one of the most common types of indirect analyses. Others are more complex but they can be reduced to algebraic expressions and solved by their aid. It should, however, be noted that the results obtained by these indirect methods cannot be depended upon for high accuracy, since slight errors in the determination of the common constituent, as chlorine in the above mixture, will cause considerable variations in the values found for the components. They should not be employed when direct methods are applicable, if accuracy is essential.

PROBLEMS

(The reactions necessary for the solution of these problems are either stated with the problem or may be found in the earlier text. In the calculations from which the answers are derived, the atomic weights given on page 195 have been employed, using, however, only the first decimal but increasing this by 1 when the second decimal is 5 or above. Thus, 39.1 has been taken as the atomic weight of potassium, 32.1 for sulphur, etc. This has been done merely to secure uniformity of treatment, and the student should remember that it is always well to take into account the degree of accuracy desired in a particular instance in determining the number of decimal places to retain. Four-place logarithms were employed in the calculations. Where four figures are given in the answer, the last figure may vary by one or (rarely) by two units, according to the method by which the problem is solved.)

VOLUMETRIC ANALYSIS

1. How many grams of pure potassium hydroxide are required for exactly 1 liter of normal alkali solution?

!Answer!: 56.1 grams.

2. Calculate the equivalent in grams (a) of sulphuric acid as an acid; (b) of hydrochloric acid as an acid; (c) of oxalic acid as an acid; (d) of nitric acid as an acid.

!Answers!: (a) 49.05; (b) 36.5; (c) 63; (d) 63.

3. Calculate the equivalent in grams of (a) potassium hydroxide; (b) of sodium carbonate; (c) of barium hydroxide; (d) of sodium bicarbonate when titrated with an acid.

!Answers!: (a) 56.1; (b) 53.8; (c) 85.7; (d) 84.

4. What is the equivalent in grams of Na_2HPO_4 (a) as a phosphate; (b) as a sodium salt?

!Answers!: (a) 47.33; (b) 71.0.

5. A sample of aqueous hydrochloric acid has a specific gravity of 1.12 and contains 23.81 per cent hydrochloric acid by weight. Calculate the grams and the milliequivalents of hydrochloric acid (HCl) in each cubic centimeter of the aqueous acid.

!Answers!: 0.2667 gram; 7.307 milliequivalents.

6. How many cubic centimeters of hydrochloric acid (sp. gr. 1.20 containing 39.80 per cent HCl by weight) are required to furnish 36.45 grams of the gaseous compound?

!Answer!: 76.33 cc.

7. A given solution contains 0.1063 equivalents of hydrochloric acid in 976 cc. What is its normal value?

!Answer!: 0.1089 N.

8. In standardizing a hydrochloric acid solution it is found that 47.26 cc. of hydrochloric acid are exactly equivalent to 1.216 grams of pure sodium carbonate, using methyl orange as an indicator. What is the normal value of the hydrochloric acid?

!Answer!: 0.4855 N.

9. Convert 42.75 cc. of 0.5162 normal hydrochloric acid to the equivalent volume of normal hydrochloric acid.

!Answer!: 22.07 cc.

10. A solution containing 25.27 cc. of 0.1065 normal hydrochloric acid is added to one containing 92.21 cc. of 0.5431 normal sulphuric acid and 50 cc. of exactly normal potassium hydroxide added from a pipette. Is the solution acid or alkaline? How many cubic centimeters of 0.1 normal acid or alkali must be added to exactly neutralize the solution?

!Answer!: 27.6 cc. alkali (solution is acid).

11. By experiment the normal value of a sulphuric acid solution is found to be 0.5172. Of this acid 39.65 cc. are exactly equivalent to 21.74 cc. of a standard alkali solution. What is the normal value of the alkali?

!Answer!: 0.9432 N.

12. A solution of sulphuric acid is standardized against a sample of calcium carbonate which has been previously accurately analyzed and found to contain 92.44% $CaCO_3$ and no other basic material. The sample weighing 0.7423 gram was titrated by adding an excess of acid (42.42 cc.) and titrating the excess with sodium hydroxide solution (11.22 cc.). 1 cc. of acid is equivalent to 0.9976 cc. of sodium hydroxide. Calculate the normal value of each.

!Answers!: Acid 0.4398 N; alkali 0.4409 N.

13. Given five 10 cc. portions of 0.1 normal hydrochloric acid, (a) how many grams of silver chloride will be precipitated by a portion when an excess of silver nitrate is added? (b) how many grams of pure anhydrous sodium carbonate (Na_2CO_3) will be neutralized by a portion of it? (c) how many grams of silver will there be in the silver chloride formed when an excess of silver nitrate is added to a portion? (d) how many grams of iron will be dissolved to $FeCl_2$ by a portion of it? (e) how many grams of magnesium chloride will be formed and how many grams of carbon dioxide liberated when an excess of magnesium carbonate is treated with a portion of the acid?

!Answers!: (a) 0.1434; (b) 0.053; (c) 0.1079; (d) 0.0279; (e) 0.04765, and 0.022.

14. If 30.00 grams of potassium tetroxalate ($KHC_2O_4 \cdot H_2C_2O_4 \cdot 2H_2O$) are dissolved and the solution diluted to exactly 1 liter, and 40 cc. are neutralized with 20 cc. of a potassium carbonate solution, what is the normal value of the carbonate solution?

!Answer!: 0.7084 N.

15. How many cubic centimeters of 0.3 normal sulphuric acid will be required to neutralize (a) 30 cc. of 0.5 normal potassium hydroxide; (b) to neutralize 30 cc. of 0.5 normal barium hydroxide; (c) to neutralize 20 cc. of a solution containing 10.02 grams of potassium bicarbonate per 100 cc.; (d) to give a precipitate of barium sulphate weighing 0.4320 gram?

!Answers!: (a) 50 cc.; (b) 50 cc.; (c) 66.73 cc.; (d) 12.33 cc.

16. It is desired to dilute a solution of sulphuric acid of which 1 cc. is equivalent to 0.1027 gram of pure sodium carbonate to make it exactly 1.250 normal. 700 cc. of the solution are available. To what volume must it be diluted?

!Answer!: 1084 cc.

17. Given the following data: 1 cc. of NaOH = 1.117 cc. HCl. The HCl is 0.4876 N. How much water must be added to 100 cc. of the alkali to make it exactly 0.5 N.?

!Answer!: 9.0 cc.

18. What is the normal value of a sulphuric acid solution which has a specific gravity of 1.839 and contains 95% H_2SO_4 by weight?

!Answer!: 35.61 N.

19. A sample of Rochelle Salt ($KNaC_4H_4O_6 \cdot 4H_2O$), after ignition in platinum to convert it to the double carbonate, is titrated with sulphuric acid, using methyl orange as an indicator. From the following data calculate the percentage purity of the sample:

Wt. sample = 0.9500 gram
H_2SO_4 used = 43.65 cc.
NaOH used = 1.72 cc.
1 cc. H_2SO_4 = 1.064 cc. NaOH
Normal value NaOH = 0.1321 N.

!Answer!: 87.72 cc.

20. One gram of a mixture of 50% sodium carbonate and 50% potassium carbonate is dissolved in water, and 17.36 cc. of 1.075 N acid is added. Is the resulting solution acid or alkaline? How many cubic centimeters of 1.075 N acid or alkali will have to be added to make the solution exactly neutral?

!Answers!: Acid; 1.86 cc. alkali.

21. In preparing an alkaline solution for use in volumetric work, an analyst, because of shortage of chemicals, mixed exactly 46.32 grams of pure KOH and 27.64 grams of pure NaOH, and after dis-

solving in water, diluted the solution to exactly one liter. How many cubic centimeters of 1.022 N hydrochloric acid are necessary to neutralize 50 cc. of the basic solution?

!Answer!: 74.18 cc.

22. One gram of crude ammonium salt is treated with strong potassium hydroxide solution. The ammonia liberated is distilled and collected in 50 cc. of 0.5 N acid and the excess titrated with 1.55 cc. of 0.5 N sodium hydroxide. Calculate the percentage of NH_3 in the sample.

!Answer!: 41.17%.

23. In titrating solutions of alkali carbonates in the presence of phenolphthalein, the color change takes place when the carbonate has been converted to bicarbonate. In the presence of methyl orange, the color change takes place only when the carbonate has been completely neutralized. From the following data, calculate the percentages of Na_2CO_3 and NaOH in an impure mixture. Weight of sample, 1.500 grams; HCl (0.5 N) required for phenolphthalein end-point, 28.85 cc.; HCl (0.5 N) required to complete the titration after adding methyl orange, 23.85 cc.

!Answers!: 6.67% NaOH; 84.28% Na_2CO_3.

24. A sample of sodium carbonate containing sodium hydroxide weighs 1.179 grams. It is titrated with 0.30 N hydrochloric acid, using phenolphthalein in cold solution as an indicator and becomes colorless after the addition of 48.16 cc. Methyl orange is added and 24.08 cc. are needed for complete neutralization. What is the percentage of NaOH and Na_2CO_3?

!Answers!: 24.50% NaOH; 64.92% Na_2CO_3.

25. From the following data, calculate the percentages of Na_2CO_3 and $NaHCO_3$ in an impure mixture. Weight of sample 1.000 gram; volume of 0.25 N hydrochloric acid required for phenolphthalein end-point, 26.40 cc.; after adding an excess of acid and boiling out the carbon dioxide, the total volume of 0.25 N hydrochloric acid required for phenolphthalein end-point, 67.10 cc.

!Answer!: 69.95% Na_2CO_3; 30.02% $NaHCO_3$.

26. In the analysis of a one-gram sample of soda ash, what must be the normality of the acid in order that the number of cubic centimeters of acid used shall represent the percentage of carbon dioxide present?

!Answer!: 0.4544 gram.

27. What weight of pearl ash must be taken for analysis in order that the number of cubic centimeters of 0.5 N acid used may be equal to one third the percentage of K_2CO_3?

!Answer!: 1.152 grams.

28. What weight of cream of tartar must have been taken for analysis in order to have obtained 97.60% $KHC_4H_4O_6$ in an analysis involving the following data: NaOH used = 30.06 cc.; H_2SO_4 solution used = 0.50 cc.; 1 cc. H_2SO_4 sol. = 0.0255 gram $CaCO_3$; 1 cc. H_2SO_4 sol. = 1.02 cc. NaOH sol.?

!Answer!: 2.846 grams.

29. Calculate the percentage of potassium oxide in an impure sample of potassium carbonate from the following data: Weight of sample = 1.00 gram; HCl sol. used = 55.90 cc.; NaOH sol. used = 0.42 cc.; 1 cc. NaOH sol. = 0.008473 gram of $KHC_2O_4 \cdot H_2C_2O_4 \cdot 2H_2O$; 2 cc. HCl sol. = 5 cc. NaOH sol.

!Answer!: 65.68%.

30. Calculate the percentage purity of a sample of calcite ($CaCO_3$) from the following data: (Standardization); Weight of $H_2C_2O_4 \cdot 2H_2O$ = 0.2460 gram; NaOH solution used = 41.03 cc.; HCl solution used = 0.63; 1 cc. NaOH solution = 1.190 cc. HCl solution. (Analysis); Weight of sample 0.1200 gram; HCl used = 36.38 cc.; NaOH used = 6.20 cc.

!Answer!: 97.97%.

31. It is desired to dilute a solution of hydrochloric acid to exactly 0.05 N. The following data are given: 44.97 cc. of the hydrochloric acid are equivalent to 43.76 cc. of the NaOH solution. The NaOH is standardized against a pure potassium tetroxalate

($KHC_2O_4 \cdot H_2C_2O_4 \cdot 2H_2O$) weighing 0.2162 gram and requires 49.14 cc. How many cc. of water must be added to 1000 cc. of the aqueous hydrochloric acid?

!Answer!: 11 cc.

32. How many cubic centimeters of 3 N phosphoric acid must be added to 300 cc. of 0.4 N phosphoric acid in order that the resulting solution may be 0.6 N?

!Answer!: 25 cc.

33. To oxidize the iron in 1 gram of $FeSO_4(NH_4)_2SO_4 \cdot 6H_2O$ (mol. wgt. 392) requires 3 cc. of a given solution of HNO_3. What is the normality of the nitric acid when used as an acid? $6FeSO_4 + 2HNO_3 + 2H_2SO_4 = 3Fe_2(SO_4)_3 + 2NO + 4H_2O$.

!Answer!: 0.2835 N.

34. The same volume of carbon dioxide at the same temperature and the same pressure is liberated from a 1 gram sample of dolomite, by adding an excess of hydrochloric acid, as can be liberated by the addition of 35 cc. of 0.5 N hydrochloric acid to an excess of any pure or impure carbonate. Calculate the percentage of CO_2 in the dolomite.

!Answer!: 38.5%.

35. How many cubic centimeters of sulphuric acid (sp. gr. 1.84, containing 96% H_2SO_4 by weight) will be required to displace the chloride in the calcium chloride formed by the action of 100 cc. of 0.1072 N hydrochloric acid on an excess of calcium carbonate, and how many grams of $CaSO_4$ will be formed?

!Answers!: 0.298 cc.; 0.7300 gram.

36. Potassium hydroxide which has been exposed to the air is found on analysis to contain 7.62% water, 2.38% K_2CO_3. and 90% KOH. What weight of residue will be obtained if one gram of this sample is added to 46 cc. of normal hydrochloric acid and the resulting solution, after exact neutralization with 1.070 N potassium hydroxide solution, is evaporated to dryness?

!Answer!: 3.47 grams.

37. A chemist received four different solutions, with the statement that they contained either pure NaOH; pure Na_2CO_3; pure $NaHCO_3$, or mixtures of these substances. From the following data identify them:

Sample I. On adding phenolphthalein to a solution of the substance, it gave no color to the solution.

Sample II. On titrating with standard acid, it required 15.26 cc. for a change in color, using phenolphthalein, and 17.90 cc. additional, using methyl orange as an indicator.

Sample III. The sample was titrated with hydrochloric acid until the pink of phenolphthalein disappeared, and on the addition of methyl orange the solution was colored pink.

Sample IV. On titrating with hydrochloric acid, using phenolphthalein, 15.00 cc. were required. A new sample of the same weight required exactly 30 cc. of the same acid for neutralization, using methyl orange.

!Answers!: (a) $NaHCO_3$; (b) $NaHCO_3 + Na_2CO_3$; (c) NaOH; (d) Na_2CO_3.

38. In the analysis of a sample of $KHC_4H_4O_6$ the following data are obtained: Weight sample = 0.4732 gram. NaOH solution used = 24.97 cc. 3.00 cc. NaOH = 1 cc. of H_3PO_4 solution of which 1 cc. will precipitate 0.01227 gram of magnesium as $MgNH_4PO_4$. Calculate the percentage of $KHC_4H_4O_6$.

!Answer!: 88.67%.

39. A one-gram sample of sodium hydroxide which has been exposed to the air for some time, is dissolved in water and diluted to exactly 500 cc. One hundred cubic centimeters of the solution, when titrated with 0.1062 N hydrochloric acid, using methyl orange as an indicator, requires 38.60 cc. for complete neutralization. Barium chloride in excess is added to a second portion of 100 cc. of the solution, which is diluted to exactly 250 cc., allowed to stand and filtered. Two hundred cubic centimeters of this filtrate require 29.62 cc. of 0.1062 N hydrochloric acid for neutralization, using phenol-

phthalein as an indicator. Calculate percentage of NaOH, Na_2CO_3, and H_2O.

!Answers!: 78.63% NaOH; 4.45% Na_2CO_3; 16.92% H_2O.

40. A sodium hydroxide solution (made from solid NaOH which has been exposed to the air) was titrated against a standard acid using methyl orange as an indicator, and was found to be exactly 0.1 N. This solution was used in the analysis of a material sold at 2 cents per pound per cent of an acid constituent A, and always mixed so that it was supposed to contain 15% of A, on the basis of the analyst's report. Owing to the carelessness of the analyst's assistant, the sodium hydroxide solution was used with phenolphthalein as an indicator in cold solution in making the analyses. The concern manufacturing this material sells 600 tons per year, and when the mistake was discovered it was estimated that at the end of a year the error in the use of indicators would either cost them or their customers $6000. Who would lose and why? Assuming the impure NaOH used originally in making the titrating solution consisted of NaOH and Na_2CO_3 only, what per cent of each was present?

!Answers!: Customer lost; 3.94% Na_2CO_3; 96.06% NaOH.

41. In the standardization of a $K_2Cr_2O_7$ solution against iron wire, 99.85% pure, 42.42 cc. of the solution were added. The weight of the wire used was 0.22 gram. 3.27 cc. of a ferrous sulphate solution having a normal value as a reducing agent of 0.1011 were added to complete the titration. Calculate the normal value of the $K_2Cr_2O_7$.

!Answer!: 0.1006 N.

42. What weight of iron ore containing 56.2% Fe should be taken to standardize an approximately 0.1 N oxidizing solution, if not more than 47 cc. are to be used?

!Answer!: 0.4667 gram.

43. One tenth gram of iron wire, 99.78% pure, is dissolved in hydrochloric acid and the iron oxidized completely with bromine water. How many grams of stannous chloride are there in a liter of solution if it requires 9.47 cc. to just reduce the iron in the above?

What is the normal value of the stannous chloride solution as a reducing agent?

!Answer!: 17.92 grams; 0.1888 N.

44. One gram of an oxide of iron is fused with potassium acid sulphate and the fusion dissolved in acid. The iron is reduced with stannous chloride, mercuric chloride is added, and the iron titrated with a normal $K_2Cr_2O_7$ solution. 12.94 cc. were used. What is the formula of the oxide, FeO, Fe_2O_3, or Fe_3O_4?

!Answer!: Fe_3O_4.

45. If an element has 98 for its atomic weight, and after reduction with stannous chloride could be oxidized by bichromate to a state corresponding to an XO_4^{-} anion, compute the oxide, or valence, corresponding to the reduced state from the following data: 0.3266 gram of the pure element, after being dissolved, was reduced with stannous chloride and oxidized by 40 cc. of $K_2Cr_2O_7$, of which one cc. = 0.1960 gram of $FeSO_4(NH_4)_2SO_4.6H_2O$.

!Answer!: Monovalent.

46. Determine the percentage of iron in a sample of limonite from the following data: Sample = 0.5000 gram. $KMnO_4$ used = 50 cc. 1 cc. $KMnO_4$ = 0.005317 gram Fe. $FeSO_4$ used = 6 cc. 1 cc. $FeSO_4$ = 0.009200 gram FeO.

!Answer!: 44.60%.

47. If 1 gram of a silicate yields 0.5000 gram of Fe_2O_3 and Al_2O_3 and the iron present requires 25 cc. of 0.2 N $KMnO_4$, calculate the percentage of FeO and Al_2O_3 in the sample.

!Answer!: 35.89% FeO; 10.03% Al_2O_3.

48. A sample of magnesia limestone has the following composition: Silica, 3.00%; ferric oxide and alumina, 0.20%; calcium oxide, 33.10%; magnesium oxide, 20.70%; carbon dioxide, 43.00%. In manufacturing lime from the above the carbon dioxide is reduced to 3.00%. How many cubic centimeters of normal $KMnO_4$ will be required to determine the calcium oxide volumetrically in a 1 gram sample of the lime?

!Answer!: 20.08 cc.

49. If 100 cc. of potassium bichromate solution (10 gram $K_2Cr_2O_7$ per liter), 5 cc. of 6 N sulphuric acid, and 75 cc. of ferrous sulphate solution (80 grams $FeSO_4.7H_2O$ per liter) are mixed, and the resulting solution titrated with 0.2121 N $KMnO_4$, how many cubic centimeters of the $KMnO_4$ solution will be required to oxidize the iron?

!Answer!: 5.70 cc.

50. If a 0.5000 gram sample of limonite containing 59.50 per cent Fe_2O_3 requires 40 cc. of $KMnO_4$ to oxidize the iron, what is the value of 1 cc. of the permanganate in terms of (a) Fe, (b) $H_2C_2O_4.2H_2O$?

!Answers!: (a) 0.005189 gram; (b) 0.005859 gram.

51. A sample of pyrolusite weighing 0.6000 gram is treated with 0.9000 gram of oxalic acid. The excess oxalic acid requires 23.95 cc. of permanganate (1 cc. = 0.03038 gram $FeSO_4.7H_2O$). What is the percentage of MnO_2, in the sample?

!Answer!: 84.47%.

52. A solution contains 50 grams of $KHC_2O_4.H_2C_2O_4.2H_2O$ per liter. What is the normal value of the solution (a) as an acid, and (b) as a reducing agent?

!Answers!: (a) 0.5903 N; (b) 0.7872 N.

53. In the analysis of an iron ore containing 60% Fe_2O_3, a sample weighing 0.5000 gram is taken and the iron is reduced with sulphurous acid. On account of failure to boil out all the excess SO_2, 38.60 cubic centimeters of 0.1 N $KMnO_4$ were required to titrate the solution. What was the error, percentage error, and what weight of sulphur dioxide was in the solution?

!Answers!: (a) 1.60%; (b) 2.67%; (c) 0.00322 gram.

54. From the following data, calculate the ratio of the nitric acid as an oxidizing agent to the tetroxalate solution as a reducing agent: 1 cc. HNO_3 = 1.246 cc. NaOH solution; 1 cc. NaOH = 1.743 cc. $KHC_2O_4.H_2C_2O_4.2H_2O$ solution; Normal value NaOH = 0.12.

!Answer!: 4.885.

55. Given the following data: 25 cc. of a hydrochloric acid, when standardized gravimetrically as silver chloride, yields a precipitate weighing 0.5465 gram. 24.35 cc. of the hydrochloric acid are exactly equivalent to 30.17 cc. of $KHC_2O_4 \cdot H_2C_2O_4 \cdot 2H_2O$ solution. How much water must be added to a liter of the oxalate solution to make it exactly 0.025 N as a reducing agent?

!Answer!: 5564 cc.

56. Ten grams of a mixture of pure potassium tetroxalate ($KHC_2O_4 \cdot H_2C_2O_4 \cdot 2H_2O$) and pure oxalic acid ($H_2C_2O_4 \cdot 2H_2O$) are dissolved in water and diluted to exactly 1000 cc. The normal value of the oxalate solution when used as an acid is 0.1315. Calculate the ratio of tetroxalate to oxalate used in making up the solution and the normal value of the solution as a reducing agent.

!Answers!: 2:1; 0.1577 N.

57. A student standardized a solution of NaOH and one of $KMnO_4$ against pure $KHC_2O_4 \cdot H_2C_2O_4 \cdot 2H_2O$ and found the former to be 0.07500 N as an alkali and the latter exactly 0.1 N as an oxidizing agent. By coincidence, exactly 47.26 cc. were used in each standardization. Find the ratio of the oxalate used in the NaOH standardization to the oxalate used in the permanganate standardization.

!Answer!: 1:1.

58. A sample of apatite weighing 0.60 gram is analyzed for its phosphoric anhydride content. If the phosphate is precipitated as $(NH_4)_3PO_4 \cdot 12MoO_3$, and the precipitate (after solution and reduction of the MoO_3 to $Mo_{24}O_{37}$), requires 100 cc. of normal $KMnO_4$ to oxidize it back to MoO_3, what is the percentage of P_2O_5?

!Answer!: 33.81%.

59. In the analysis of a sample of steel weighing 1.881 grams the phosphorus was precipitated with ammonium molybdate and the yellow precipitate was dissolved, reduced and titrated with $KMnO_4$. If the sample contained 0.025 per cent P and 6.01 cc. of

$KMnO_4$ were used, to what oxide was the molybdenum reduced? 1 cc. $KMnO_4$ = 0.007188 gram $Na_2C_2O_4$.

!Answer!: Mo_4O_5.

60. What is the value of 1 cc. of an iodine solution (1 cc. equivalent to 0.0300 gram $Na_2S_2O_3$) in terms of As_2O_3?

!Answer!: 0.009385 gram.

61. 48 cc. of a solution of sodium thiosulphate are required to titrate the iodine liberated from an excess of potassium iodide solution by 0.3000 gram of pure KIO_3. (KIO_3 + 5KI + $3H_2SO_4$ = $3K_2SO_4$ + $3I_2$ + $3H_2O$.) What is the normal strength of the sodium thiosulphate and the value of 1 cc. of it in terms of iodine?

!Answers!: 0.1753 N; 0.02224 gram.

62. One thousand cubic centimeters of 0.1079 N sodium thiosulphate solution is allowed to stand. One per cent by weight of the thiosulphate is decomposed by the carbonic acid present in the solution. To what volume must the solution be diluted to make it exactly 0.1 N as a reducing agent? ($Na_2S_2O_3$ + $2H_2CO_3$ = H_2SO_3 + $2NaHCO_3$ + S.)

!Answer!: 1090 cc.

63. An analyzed sample of stibnite containing 70.05% Sb is given for analysis. A student titrates it with a solution of iodine of which 1 cc. is equivalent to 0.004950 gram of As_2O_3. Due to an error on his part in standardization, the student's analysis shows the sample to contain 70.32% Sb. Calculate the true normal value of the iodine solution, and the percentage error in the analysis.

!Answers!: 0.1000 N; 0.39%.

64. A sample of pyrolusite weighing 0.5000 gram is treated with an excess of hydrochloric acid, the liberated chlorine is passed into potassium iodide and the liberated iodine is titrated with sodium thiosulphate solution (49.66 grams of pure $Na_2S_2O_3.5H_2O$ per liter). If 38.72 cc. are required, what volume of 0.25 normal permanganate solution will be required in an indirect determination in which a similar sample is reduced with

0.9012 gram $H_2C_2O_4 \cdot 2H_2O$ and the excess oxalic acid titrated?

!Answer!: 26.22 cc.

65. In the determination of sulphur in steel by evolving the sulphur as hydrogen sulphide, precipitating cadmium sulphide by passing the liberated hydrogen sulphide through ammoniacal cadmium chloride solution, and decomposing the CdS with acid in the presence of a measured amount of standard iodine, the following data are obtained: Sample, 5.027 grams; cc. $Na_2S_2O_3$ sol. = 12.68; cc. Iodine sol. = 15.59; 1 cc. Iodine sol. = 1.086 cc. $Na_2S_2O_3$ sol.; 1 cc. $Na_2S_2O_3$ = 0.005044 gram Cu. Calculate the percentage of sulphur. ($H_2S + I_2 = 2HI + S$.)

!Answer!: 0.107%.

66. Given the following data, calculate the percentage of iron in a sample of crude ferric chloride weighing 1.000 gram. The iodine liberated by the reaction $2FeCl_3 + 2HI = 2HCl + 2FeCl_2 + I_2$ is reduced by the addition of 50 cc. of sodium thiosulphate solution and the excess thiosulphate is titrated with standard iodine and requires 7.85 cc. 45 cc. I_2 solution = 45.95 cc. $Na_2S_2O_3$ solution; 45 cc. As_2O_3 solution = 45.27 cc. I_2 solution. 1 cc. arsenite solution = 0.005160 gram As_2O_3.

!Answer!: 23.77%.

67. Sulphide sulphur was determined in a sample of reduced barium sulphate by the evolution method, in which the sulphur was evolved as hydrogen sulphide and was passed into $CdCl_2$ solution, the acidified precipitate being titrated with iodine and thiosulphate. Sample, 5.076 grams; cc. I_2 = 20.83; cc. $Na_2S_2O_3$ = 12.37; 43.45 cc. $Na_2S_2O_3$ = 43.42 cc. I_2; 8.06 cc. $KMnO_4$ = 44.66 cc. $Na_2S_2O_3$; 28.87 cc. $KMnO_4$ = 0.2004 gram $Na_2C_2O_4$. Calculate the percentage of sulphide sulphur in the sample.

!Answer!: 0.050%.

68. What weight of pyrolusite containing 89.21% MnO_2 will oxidize the same amount of oxalic acid as 37.12 cc. of a permanga-

nate solution, of which 1 cc. will liberate 0.0175 gram of I_2 from KI?

!Answer!: 0.2493 gram.

69. A sample of pyrolusite weighs 0.2400 gram and is 92.50% pure MnO_2. The iodine liberated from KI by the manganese dioxide is sufficient to react with 46.24 cc. of $Na_2S_2O_3$ sol. What is the normal value of the thiosulphate?

!Answer!:: 0.1105 N.

70. In the volumetric analysis of silver coin (90% Ag), using a 0.5000 gram sample, what is the least normal value that a potassium thiocyanate solution may have and not require more than 50 cc. of solution in the analysis?

!Answer!: 0.08339 N.

71. A mixture of pure lithium chloride and barium bromide weighing 0.6 gram is treated with 45.15 cubic centimeters of 0.2017 N silver nitrate, and the excess titrated with 25 cc. of 0.1 N KSCN solution, using ferric alum as an indicator. Calculate the percentage of bromine in the sample.

!Answer!: 40.11%.

72. A mixture of the chlorides of sodium and potassium from 0.5000 gram of a feldspar weighs 0.1500 gram, and after solution in water requires 22.71 cc. of 0.1012 N silver nitrate for the precipitation of the chloride ions. What are the percentages of Na_2O and K_2O in the feldspar?

!Answer!: 8.24% Na_2O; 9.14% K_2O.

GRAVIMETRIC ANALYSIS

73. Calculate (a) the grams of silver in one gram of silver chloride; (b) the grams of carbon dioxide liberated by the addition of an excess of acid to one gram of calcium carbonate; (c) the grams of $MgCl_2$ necessary to precipitate 1 gram of $MgNH_4PO_4$.

!Answers!: (a) 0.7526; (b) 0.4397; (c) 0.6940.

74. Calculate the chemical factor for (a) Sn in SnO_2; (b) MgO in $Mg_2P_2O_7$; (c) P_2O_5 in $Mg_2P_2O_7$; (d) Fe in Fe_2O_3; (e) SO_4 in $BaSO_4$.

!Answers!: (a) 0.7879; (b) 0.3620; (c) 0.6378; (d) 0.6990; (e) 0.4115.

75. Calculate the log factor for (a) Pb in $PbCrO_4$; (b) Cr_2O_3 in $PbCrO_4$; (c) Pb in PbO_2 and (d) CaO in CaC_2O_4.

!Answers!: (a) 9.8069-10, (b) 9.3713-10; (c) 9.9376-10; (d) 9.6415-10.

76. How many grams of Mn_3O_4 can be obtained from 1 gram of MnO_2?

!Answer!: 0.8774 gram.

77. If a sample of silver coin weighing 0.2500 gram gives a precipitate of AgCl weighing 0.2991 gram, what weight of AgI could have been obtained from the same weight of sample, and what is the percentage of silver in the coin?

!Answers!: 0.4898 gr.; 90.05%.

78. How many cubic centimeters of hydrochloric acid (sp. gr. 1.13 containing 25.75% HCl by weight) are required to exactly neutralize 25 cc. of ammonium hydroxide (sp. gr. .90 containing 28.33% NH_3 by weight)?

!Answer!: 47.03 cc.

79. How many cubic centimeters of ammonium hydroxide solution (sp. gr. 0.96 containing 9.91% NH_3 by weight) are required to precipitate the aluminium as aluminium hydroxide from a two-gram sample of alum $(KAl(SO_4)_2.12H_2O)$? What will be the weight of the ignited precipitate?

!Answers!: 2.26 cc.; 0.2154 gram.

80. What volume of nitric acid (sp. gr. 1.05 containing 9.0% HNO_3 by weight) is required to oxidize the iron in one gram of $FeSO_4.7H_2O$ in the presence of sulphuric acid? $6FeSO_4 + 2HNO_3 + 3H_2SO_4 = 3Fe_2(SO_4)_3 + 2NO + 4H_2O$.

!Answer!: 0.80 cc.

81. If 0.7530 gram of ferric nitrate ($Fe(NO_3)_3 \cdot 9H_2O$) is dissolved in water and 1.37 cc. of HCl (sp. gr. 1.11 containing 21.92% HCl by weight) is added, how many cubic centimeters of ammonia (sp. gr. 0.96 containing 9.91% NH_3 by weight) are required to neutralize the acid and precipitate the iron as ferric hydroxide?

!Answer!: 2.63 cc.

82. To a suspension of 0.3100 gram of $Al(OH)_3$ in water are added 13.00 cc. of aqueous ammonia (sp. gr. 0.90 containing 28.4% NH_3 by weight). How many cubic centimeters of sulphuric acid (sp. gr. 1.18 containing 24.7% H_2SO_4 by weight) must be added to the mixture in order to bring the aluminium into solution?

!Answer!: 34.8 cc.

83. How many cubic centimeters of sulphurous acid (sp. gr. 1.04 containing 75 grams SO_2 per liter) are required to reduce the iron in 1 gram of ferric alum ($KFe(SO_4)_2 \cdot 12H_2O$)? $Fe_2(SO_4)_3 + SO_2 + 2H_2O = 2FeSO_4 + 2H_2SO_4$.

!Answer!: 0.85 cc.

84. How many cubic centimeters of a solution of potassium bichromate containing 26.30 grams of $K_2Cr_2O_7$ per liter must be taken in order to yield 0.6033 gram of Cr_2O_3 after reduction and precipitation of the chromium?

$K_2Cr_2O_7 + 3SO_2 + H_2SO_4 = K_2SO_4 + Cr_2(SO_4)_3 + H_2O$.

!Answer!: 44.39 cc.

85. How many cubic centimeters of ammonium hydroxide (sp. gr. 0.946 containing 13.88% NH_3 by weight) are required to precipitate the iron as $Fe(OH)_3$ from a sample of pure $FeSO_4 \cdot (NH_4)_2SO_4 \cdot 6H_2O$, which requires 0.34 cc. of nitric acid (sp. gr. 1.350 containing 55.79% HNO_3 by weight) for oxidation of the iron? (See problem No. 80 for reaction.)

!Answer!: 4.74 cc.

86. In the analysis of an iron ore by solution, oxidation and precipitation of the iron as $Fe(OH)_3$, what weight of sample must be taken

for analysis so that each one hundredth of a gram of the ignited precipitate of Fe_2O_3 shall represent one tenth of one per cent of iron?

!Answer!: 6.99 grams.

87. What weight in grams of impure ferrous ammonium sulphate should be taken for analysis so that the number of centigrams of $BaSO_4$ obtained will represent five times the percentage of sulphur in the sample?

!Answer!: 0.6870 gram.

88. What weight of magnetite must be taken for analysis in order that, after precipitating and igniting all the iron to Fe_2O_3, the percentage of Fe_2O_4 in the sample may be found by multiplying the weight in grams of the ignited precipitate by 100?

!Answer!: 0.9665 gram.

89. After oxidizing the arsenic in 0.5000 gram of pure As_2S_3 to arsenic acid, it is precipitated with "magnesia mixture" ($MgCl_2$ + $2NH_4Cl$). If exactly 12.6 cc. of the mixture are required, how many grams of $MgCl_2$ per liter does the solution contain? $H_3AsO_4 + MgCl_2 + 3NH_4OH = MgNH_4AsO_4 + 2NH_4Cl + 3H_2O$.

!Answer!: 30.71 grams.

90. A sample is prepared for student analysis by mixing pure apatite ($Ca_3(PO_4)_2.CaCl_2$) with an inert material. If 1 gram of the sample gives 0.4013 gram of $Mg_2P_2O_7$, how many cubic centimeters of ammonium oxalate solution (containing 40 grams of $(NH_4)_2C_2O_4.H_2O$ per liter) would be required to precipitate the calcium from the same weight of sample?

!Answer!: 25.60 cc.

91. If 0.6742 gram of a mixture of pure magnesium carbonate and pure calcium carbonate, when treated with an excess of hydrochloric acid, yields 0.3117 gram of carbon dioxide, calculate the percentage of magnesium oxide and of calcium oxide in the sample.

!Answers!: 13.22% MgO; 40.54% CaO. 92. The calcium in a sample of dolomite weighing 0.9380 gram is precipitated as calcium oxalate and ignited to calcium oxide. What volume of gas, measured over water at 20°C. and 765 mm. pressure, is given off during ignition, if the resulting oxide weighs 0.2606 gram? (G.M.V. = 22.4 liters; V.P. water at 20°C. = 17.4 mm.)

!Answer!: 227 cc.

93. A limestone is found to contain 93.05% $CaCO_3$, and 5.16 % $MgCO_3$. Calculate the weight of CaO obtainable from 3 tons of the limestone, assuming complete conversion to oxide. What weight of $Mg_2P_2O_7$ could be obtained from a 3-gram sample of the limestone?

!Answers!: 1.565 tons; 0.2044 gram.

94. A sample of dolomite is analyzed for calcium by precipitating as the oxalate and igniting the precipitate. The ignited product is assumed to be CaO and the analyst reports 29.50% Ca in the sample. Owing to insufficient ignition, the product actually contained 8% of its weight of $CaCO_3$. What is the correct percentage of calcium in the sample, and what is the percentage error?

!Answers!: 28.46%; 3.65% error.

95. What weight of impure calcite ($CaCO_3$) should be taken for analysis so that the volume in cubic centimeters of CO_2 obtained by treating with acid, measured dry at 18°C. and 763 mm., shall equal the percentage of CaO in the sample?

!Answer!: 0.2359 gram.

96. How many cubic centimeters of HNO_3 (sp. gr. 1.13 containing 21.0% HNO_3 by weight) are required to dissolve 5 grams of brass, containing 0.61% Pb, 24.39% Zn, and 75% Cu, assuming reduction of the nitric acid to NO by each constituent? What fraction of this volume of acid is used for oxidation?

!Answers!: 55.06 cc.; 25%.

97. What weight of metallic copper will be deposited from a cupric salt solution by a current of 1.5 amperes during a period of 45 minutes, assuming 100% current efficiency? (1 Faraday = 96,500 coulombs.)

!Answer!: 1.335 grams.

98. In the electrolysis of a 0.8000 gram sample of brass, there is obtained 0.0030 gram of PbO_2, and a deposit of metallic copper exactly equal in weight to the ignited precipitate of $Zn_2P_2O_7$ subsequently obtained from the solution. What is the percentage composition of the brass?

!Answers!: 69.75% Cu; 29.92% Zn; 0.33% Pb.

99. A sample of brass (68.90% Cu; 1.10% Pb and 30.00% Zn) weighing 0.9400 gram is dissolved in nitric acid. The lead is determined by weighing as $PbSO_4$, the copper by electrolysis and the zinc by precipitation with $(NH_4)_2HPO_4$ in a neutral solution.

(a) Calculate the cubic centimeters of nitric acid (sp. gr. 1.42 containing 69.90% HNO_3 by weight) required to just dissolve the brass, assuming reduction to NO.

!Answer!: 2.48 cc.

(b) Calculate the cubic centimeters of sulphuric acid (sp. gr. 1.84 containing 94% H_2SO_4 by weight) to displace the nitric acid.

!Answer!: 0.83 cc.

(c) Calculate the weight of $PbSO_4$.

!Answer!: 0.0152 gram.

(d) The clean electrode weighs 10.9640 grams. Calculate the weight after the copper has been deposited.

!Answer!: 11.6116 grams.

(e) Calculate the grams of $(NH_4)_2HPO_4$ required to precipitate the zinc as $ZnNH_4PO_4$.

!Answer!: 0.5705 gram.

(f) Calculate the weight of ignited $Zn_2P_2O_7$.

!Answer!: 0.6573 gram.

100. If in the analysis of a brass containing 28.00% zinc an error is made in weighing a 2.5 gram portion by which 0.001 gram too much is weighed out, what percentage error in the zinc determination would result? What volume of a solution of sodium hydrogen phosphate, containing 90 grams of $Na_2HPO_4 \cdot 12H_2O$ per

liter, would be required to precipitate the zinc as $ZnNH_4PO_4$ and what weight of precipitate would be obtained?

!Answers!: (a) 0.04% error; (b) 39.97 cc.; (c) 1.909 grams.

101. A sample of magnesium carbonate, contaminated with SiO_2 as its only impurity, weighs 0.5000 gram and loses 0.1000 gram on ignition. What volume of disodium phosphate solution (containing 90 grams $Na_2HPO_4 \cdot 12H_2O$ per liter) will be required to precipitate the magnesium as magnesium ammonium phosphate?

!Answer!: 9.07 cc.

102. 2.62 cubic centimeters of nitric acid (sp. gr. 1.42 containing 69.80% HNO_2 by weight) are required to just dissolve a sample of brass containing 69.27% Cu; 0.05% Pb; 0.07% Fe; and 30.61% Zn. Assuming the acid used as oxidizing agent was reduced to NO in every case, calculate the weight of the brass and the cubic centimeters of acid used as acid.

!Answer!: 0.992 gram; 1.97 cc.

103. One gram of a mixture of silver chloride and silver bromide is found to contain 0.6635 gram of silver. What is the percentage of bromine?

!Answer!: 21.30%.

104. A precipitate of silver chloride and silver bromide weighs 0.8132 gram. On heating in a current of chlorine, the silver bromide is converted to silver chloride, and the mixture loses 0.1450 gram in weight. Calculate the percentage of chlorine in the original precipitate.

!Answer!: 6.13%.

105. A sample of feldspar weighing 1.000 gram is fused and the silica determined. The weight of silica is 0.6460 gram. This is fused with 4 grams of sodium carbonate. How many grams of the carbonate actually combined with the silica in fusion, and what was the loss in weight due to carbon dioxide during the fusion?

!Answers!: 1.135 grams; 0.4715 gram.

106. A mixture of barium oxide and calcium oxide weighing 2.2120 grams is transformed into mixed sulphates, weighing 5.023 grams. Calculate the grams of calcium oxide and barium oxide in the mixture.

!Answers!: 1.824 grams CaO; 0.3877 gram BaO.

APPENDIX

ELECTROLYTIC DISSOCIATION THEORY

The following brief statements concerning the ionic theory and a few of its applications are intended for reference in connection with the explanations which are given in the Notes accompanying the various procedures. The reader who desires a more extended discussion of the fundamental theory and its uses is referred to such books as Talbot and Blanchard's !Electrolytic Dissociation Theory! (Macmillan Company), or Alexander Smith's !Introduction to General Inorganic Chemistry! (Century Company).

The !electrolytic dissociation theory!, as propounded by Arrhenius in 1887, assumes that acids, bases, and salts (that is, electrolytes) in aqueous solution are dissociated to a greater or less extent into !ions!. These ions are assumed to be electrically charged atoms or groups of atoms, as, for example, H^{+} and Br^{-} from hydrobromic acid, Na^{+} and OH^{-} from sodium hydroxide, $2NH_{4}^{+}$ and SO_{4}^{-} from ammonium sulphate. The unit charge is that which is dissociated with a hydrogen ion. Those upon other ions vary in sign and number according to the chemical character and valence of the atoms or radicals of which the ions are composed. In any solution the aggregate of the positive charges upon the positive ions (!cations!) must always balance the aggregate negative charges upon the negative ions (!anions!).

It is assumed that the Na^{+} ion, for example, differs from the sodium atom in behavior because of the very considerable electrical charge which it carries and which, as just stated, must, in an electrically neutral solution, be balanced by a corresponding negative charge on some other ion. When an electric current is passed through a solution of an electrolyte the ions move with and convey the current, and when the cations come into contact with the negatively charged cathode they lose their charges, and the resulting electrically neutral atoms (or radicals) are liberated as such, or else enter at once into chemical reaction with the components of the solution.

Two ions of identically the same composition but with different electrical charges may exhibit widely different properties. For example, the ion MnO_4^{-} from permanganates yields a purple-red solution and differs in its chemical behavior from the ion $MnO_4^{=}$ from manganates, the solutions of which are green.

The chemical changes upon which the procedures of analytical chemistry depend are almost exclusively those in which the reacting substances are electrolytes, and analytical chemistry is, therefore, essentially the chemistry of the ions. The percentage dissociation of the same electrolyte tends to increase with increasing dilution of its solution, although not in direct proportion. The percentage dissociation of different electrolytes in solutions of equivalent concentrations (such, for example, as normal solutions) varies widely, as is indicated in the following tables, in which approximate figures are given for tenth-normal solutions at a temperature of about 18°C.

ACIDS

SUBSTANCE	PERCENTAGE DISSOCIATION IN 0.1 EQUIVALENT SOLUTION
HCl, HBr, HI, HNO_3	90
$HClO_3$, $HClO_4$, $HMnO_4$	90
$H_2SO_4 \longleftrightarrow H^{+} + HSO_4^{-}$	90
$H_2C_2O_4 \longleftrightarrow H^{+} + HC_2O_4^{-}$	50
$H_2SO_3 \longleftrightarrow H^{+} + HSO_3^{-}$	20
$H_3PO_4 \longleftrightarrow H^{+} + H_2PO_4^{-}$	27

$H_2PO_4^- <\!-\!> H^+ + HPO_4^{2-}$	0.2
$H_3AsO_4 <\!-\!> H^+ + H_2AsO_4^-$	20
HF	9
$HC_2H_3O_2$	1.4
$H_2CO_3 <\!-\!> H^+ + HCO_3^-$	0.12
$H_2S <\!-\!> H^+ + HS^-$	0.05
HCN	0.01

BASES

SUBSTANCE	PERCENTAGE DISSOCIATION IN 0.1 EQUIVALENT SOLUTION
KOH, NaOH	86
$Ba(OH)_2$	75
NH_4OH	1.4

SALTS

TYPE OF SALT	PERCENTAGE DISSOCIATION IN 0.1 EQUIVALENT SOLUTION
R^+R^-	86
$R^{++}(R^-)_2$	72
$(R^+)_2 R^-$	72
$R^{++}R^{--}$	45

The percentage dissociation is determined by studying the electrical conductivity of the solutions and by other physico-chemical methods, and the following general statements summarize the results:

!Salts!, as a class, are largely dissociated in aqueous solution.

!Acids! yield H^+ ions in water solution, and the comparative !strength!, that is, the activity, of acids is proportional to the concentration of the H^+ ions and is measured by the percentage dissociation in solutions of equivalent concentration. The common mineral acids are largely dissociated and therefore give a relatively high concentration of H^+ ions, and are commonly known as "strong acids." The organic acids, on the other hand, belong generally to the group of "weak acids."

!Bases! yield OH^- ions in water solution, and the comparative strength of the bases is measured by their relative dissociation in solutions of equivalent concentration. Ammonium hydroxide is a weak base, as shown in the table above, while the hydroxides of sodium and potassium exhibit strongly basic properties.

Ionic reactions are all, to a greater or less degree, !reversible reactions!. A typical example of an easily reversible reaction is that representing the changes in ionization which an electrolyte such as acetic acid undergoes on dilution or concentration of its solutions, !i.e.!, $HC_2H_3O_2 \longleftrightarrow H^+ + C_2H_3O_2^-$. As was stated above, the ionization increases with dilution, the reaction then proceeding from left to right, while concentration of the solution occasions a partial reassociation of the ions, and the reaction proceeds from right to left. To understand the principle underlying these changes it is necessary to consider first the conditions which prevail when a solution of acetic acid, which has been stirred until it is of uniform concentration throughout, has come to a constant temperature. A careful study of such solutions has shown that there is a definite state of equilibrium between the constituents of the solution; that is, there is a definite relation between the undissociated acetic acid and its ions, which is characteristic for the prevailing conditions. It is not, however, assumed that this is a condition of static equilibrium, but rather that there is continual dissociation and association, as represented by the opposing reactions, the apparent condition of rest resulting from the fact that the amount of change in one direction during a given time is exactly equal to that in the opposite direction. A quantitative study of the amount of undissociated acid, and of H^+ ions and $C_2H_3O_2^-$ ions actually to be found in a large number of solutions of acetic acid of varying dilution (assuming them to be in a condition of equilibrium at a common temperature), has shown that there is always a definite relation between these three quantities which may be expressed thus:

(!Conc'n H^+ x Conc'n $C_2H_3O_2^-$)/Conc'n $HC_2H_3O_2$ = Constant!.

In other words, there is always a definite and constant ratio between the product of the concentrations of the ions and the concentration of the undissociated acid when conditions of equilibrium prevail.

It has been found, further, that a similar statement may be made regarding all reversible reactions, which may be expressed in general terms thus: The rate of chemical change is proportional to the product of the concentrations of the substances taking part in the reaction; or, if conditions of equilibrium are considered in which, as stated, the rate of change in opposite directions is assumed to be equal, then the product of the concentrations of the substances entering into the reaction stands in a constant ratio to the product of the concentrations of the resulting substances, as given in the expression above for the solutions of acetic acid. This principle is called the !Law of Mass Action!.

It should be borne in mind that the expression above for acetic acid applies to a wide range of dilutions, provided the temperature remains constant. If the temperature changes the value of the constant changes somewhat, but is again uniform for different dilutions at that temperature. The following data are given for temperatures of about 18°C.[1]

MOLAL CONCENTRATION	FRACTION IONIZED	MOLAL CONCENTRATION OF H^{+} AND ACETATE^{-} IONS	MOLAL CONCENTRATION OF UNDISSOCIATED ACID	VALUE OF CONSTANT
1.0	.004	.004	.996	.0000161
0.1	.013	.0013	.0987	.0000171
0.01	.0407	.000407	.009593	.0000172

==========================

[Footnote 1: Alexander Smith, !General Inorganic Chemistry!, p. 579.]

The molal concentrations given in the table refer to fractions of a gram-molecule per liter of the undissociated acid, and to fractions of the corresponding quantities of H^{+} and $C_2H_3O_2^{-}$ ions per liter which would result from the complete dissociation of a gram-molecule of acetic acid. The values calculated for the constant are subject to some variation on account of experimental errors in determining the percentage ionized in each case, but the approximate agreement between the values found for molal and centimolal (one hundredfold dilution) is significant.

The figures given also illustrate the general principle, that the !relative! ionization of an electrolyte increases with the dilution of its solution. If we consider what happens during the (usually) brief period of dilution of the solution from molal to 0.1 molal, for example, it will be seen that on the addition of water the conditions of concentration which led to equality in the rate of change, and hence to equilibrium in the molal solution, cease to exist; and since the dissociating tendency increases with dilution, as just stated, it is true at the first instant after the addition of water that the concentration of the undissociated acid is too great to be permanent under the new conditions of dilution, and the reaction, $HC_2H_3O_2 <\!-\!> H^{+} + C_2H_3O_2^{-}$, will proceed from left to right with great rapidity until the respective concentrations adjust themselves to the new conditions.

That which is true of this reaction is also true of all reversible reactions, namely, that any change of conditions which occasions an increase or a decrease in concentration of one or more of the components causes the reaction to proceed in one direction or the other until a new state of equilibrium is established. This principle is constantly applied throughout the discussion of the applications of the ionic theory in analytical chemistry, and it should be clearly understood that whenever an existing state of equilibrium is disturbed as a result of changes of dilution or temperature, or as a consequence of chemical changes which bring into action any of the constituents of the solution, thus altering their concentrations, there is always a tendency to re-establish this equilibrium in accordance with the law. Thus, if a base is added to the solution of acetic acid the H^+ ions then unite with the OH^- ions from the base to form undissociated water. The concentration of the H^+ ions is thus diminished, and more of the acid dissociates in an attempt to restore equilbrium, until finally practically all the acid is dissociated and neutralized.

Similar conditions prevail when, for example, silver ions react with chloride ions, or barium ions react with sulphate ions. In the former case the dissociation reaction of the silver nitrate is $AgNO_3 \longleftrightarrow Ag^+ + NO_3^-$, and as soon as the Ag^+ ions unite with the Cl^- ions the concentration of the former is diminished, more of the $AgNO_3$ dissociates, and this process goes on until the Ag^+ ions are practically all removed from the solution, if the Cl^- ions are present in sufficient quantity.

For the sake of accuracy it should be stated that the mass law cannot be rigidly applied to solutions of those electrolytes which are largely dissociated. While the explanation of the deviation from quantitative exactness in these cases is not known, the law is still of marked service in developing analytical methods along more logical lines than was formerly practicable. It has not seemed wise to qualify each statement made in the Notes to indicate this lack of quantitative exactness. The student should recognize its existence, however, and will realize its significance better as his knowledge of physical chemistry increases.

If we apply the mass law to the case of a substance of small solubility, such as the compounds usually precipitated in quantitative analysis, we derive what is known as the !solubility product!, as follows: Taking silver chloride as an example, and remembering that it is not absolutely insoluble in water, the equilibrium expression for its solution is:

(!Conc'n Ag^+ x Conc'n Cl^-)/Conc'n AgCl = Constant!.

But such a solution of silver chloride which is in contact with the solid precipitate must be saturated for the existing temperature, and the quantity of undissociated AgCl in the solution is definite and constant for that temperature. Since it is a constant, it may be eliminated, and the expression becomes [Conc'n Ag^{+} x Conc'n Cl^{-} = Constant], and this is known as the solubility product. No precipitation of a specific substance will occur until the product of the concentrations of its ions in a solution exceeds the solubility product for that substance; whenever that product is exceeded precipitation must follow.

It will readily be seen that if a substance which yields an ion in common with the precipitated compound is added to such a solution as has just been described, the concentration of that ion is increased, and as a result the concentration of the other ion must proportionately decrease, which can only occur through the formation of some of the undissociated compound which must separate from the already saturated solution. This explains why the addition of an excess of the precipitant is often advantageous in quantitative procedures. Such a case is discussed at length in Note 2 on page 113.

Similarly, the ionization of a specific substance in solution tends to diminish on the addition of another substance with a common ion, as, for instance, the addition of hydrochloric acid to a solution of hydrogen sulphide. Hydrogen sulphide is a weak acid, and the concentration of the hydrogen ions in its aqueous solutions is very small. The equilibrium in such a solution may be represented as:

$([Conc'n H^{+}])^{2}$ x Conc'n $S^{-})/$Conc'n $H_{2}S$ = Constant],
and a marked increase in the concentration of the H^{+} ions, such as would result from the addition of even a small amount of the highly ionized hydrochloric acid, displaces the point of equilibrium and some of the S^{-} ions unite with H^{+} ions to form undissociated $H_{2}S$. This is of much importance in studying the reactions in which hydrogen sulphide is employed, as in qualitative analysis. By a parallel course of reasoning it will be seen that the addition of a salt of a weak acid or base to solutions of that acid or base make it, in effect, still weaker because they decrease its percentage ionization.

To understand the changes which occur when solids are dissolved where chemical action is involved, it should be remembered that no substance is completely insoluble in water, and that those products of a chemical change which are least dissociated will first form. Consider, for example, the action of hydrochloric acid upon magnesium hydroxide. The minute quantity of dissolved hydroxide dissociates thus: $Mg(OH)_2 \longleftrightarrow Mg^{++} + 2OH^{-}$. When the acid is introduced, the H^{+} ions of the acid unite with the OH^{-} ions to form undissociated water. The concentration of the OH^{-} ions is thus diminished, more $Mg(OH)_2$ dissociates, the solution is no longer saturated with the undissociated compound, and more of the solid dissolves. This process repeats itself with great rapidity until, if sufficient acid is present, the solid passes completely into solution.

Exactly the same sort of process takes place if calcium oxalate, for example, is dissolved in hydrochloric acid. The $C_2O_4^{--}$ ions unite with the H^{+} ions to form undissociated oxalic acid, the acid being less dissociated than normally in the presence of the H^{+} ions from the hydrochloric acid (see statements regarding hydrogen sulphide above). As the undissociated oxalic acid forms, the concentration of the $C_2O_4^{--}$ ions lessens and more CaC_2O_4 dissolves, as described for the $Mg(OH)_2$ above. Numerous instances of the applications of these principles are given in the Notes.

Water itself is slightly dissociated, and although the resulting H^{+} and OH^{-} ions are present only in minute concentrations (1 mol. of dissociated water in 10^{7} liters), yet under some conditions they may give rise to important consequences. The term !hydrolysis! is applied to the changes which result from the reaction of these ions. Any salt which is derived from a weak base or a weak acid (or both) is subject to hydrolytic action. Potassium cyanide, for example, when dissolved in water gives an alkaline solution because some of the H^{+} ions from the water unite with CN^{-} ions to form (HCN), which is a very weak acid, and is but very slightly dissociated. Potassium hydroxide, which might form from the OH^{-} ions, is so largely dissociated that the OH^{-} ions remain as such in the solution. The union of the H^{+} ions with the CN^{-} ions to form the undissociated HCN diminishes the concentration of the H^{+} ions, and more water dissociates ($H_2O \longleftrightarrow H^{+} + OH^{-}$) to restore the equilibrium. It is clear, however, that there must be a gradual accumulation of OH^{-} ions in the solution as a result of these changes, causing the solution to exhibit an alkaline reaction, and also that ultimately the further dissociation of the water will be checked by the presence of these ions, just as the dissociation of the H_2S was lessened by the addition of HCl.

An exactly opposite result follows the solution of such a salt as $Al_2(SO_4)_3$ in water. *In this case the acid is strong and the base weak, and the OH^- ions form the little dissociated $Al(OH)_3$*, while the H^+ ions remain as such in the solution, sulphuric acid being extensively dissociated. The solution exhibits an acid reaction.

Such hydrolytic processes as the above are of great importance in analytical chemistry, especially in the understanding of the action of indicators in volumetric analysis. (See page 32.)

The impelling force which causes an element to pass from the atomic to the ionic condition is termed !electrolytic solution pressure!, or ionization tension. This force may be measured in terms of electrical potential, and the table below shows the relative values for a number of elements.

In general, an element with a greater solution pressure tends to cause the deposition of an element of less solution pressure when placed in a solution of its salt, as, for instance, when a strip of zinc or iron is placed in a solution of a copper salt, with the resulting precipitation of metallic copper.

Hydrogen is included in the table, and its position should be noted with reference to the other common elements. For a more extended discussion of this topic the student should refer to other treatises.

POTENTIAL SERIES OF THE METALS

			POTENTIAL		POTENTIAL	IN VOLTS		IN VOLTS
			Sodium Na^+	+2.44	Lead Pb^{++}	-0.13	Calcium Ca^{++}	
	Hydrogen H^+	-0.28	Magnesium Mg^{++}		Bismuth Bi^{+++}		Aluminum Al^{+++}	+1.00
Antimony	-0.75	Manganese Mn^{++}			Arsenic		Zinc Zn^{++}	+0.49
Copper Cu^{++}	-0.61	Cadmium Cd^{++}	+0.14	Mercury Hg^+	-1.03	Iron Fe^{++}	+0.063	Silver Ag^+	-1.05 Cobalt

| Co^{++} | -0.045 | Platinum | Nickel Ni^{++} | -0.049 | Gold | Tin Sn^{++} | | -0.085(?) | | |

THE FOLDING OF A FILTER PAPER

If a filter paper is folded along its diameter, and again folded along the radius at right angles to the original fold, a cone is formed on opening, the angle of which is 60°. Funnels for analytical use are supposed to have the same angle, but are rarely accurate. It is possible, however, with care, to fit a filter thus folded into a funnel in such a way as to prevent air from passing down between the paper and the funnel to break the column of liquid in the stem, which aids greatly, by its gentle suction, in promoting the rate of filtration.

Such a filter has, however, the disadvantage that there are three thicknesses of paper back of half of its filtering surface, as a consequence of which one half of a precipitate washes or drains more slowly. Much time may be saved in the aggregate by learning to fold a filter in such a way as to improve its effective filtering surface. The directions which follow, though apparently complicated on first reading, are easily applied and easily remembered. Use a 6-inch filter for practice. Place four dots on the filter, two each on diameters which are at right angles to each other. Then proceed as follows: (1) Fold the filter evenly across one of the diameters, creasing it carefully; (2) open the paper, turn it over, rotate it 90° to the right, bring the edges together and crease along the other diameter; (3) open, and rotate 45° to the right, bring edges together, and crease evenly; (4) open, and rotate 90° to the right, and crease evenly; (5) open, turn the filter over, rotate 22-(1/2)° to the right, and crease evenly; (6) open, rotate 45° to the right and crease evenly; (7) open, rotate 45° to the right and crease evenly; (8) open, rotate 45° to the right and crease evenly; (9) open the filter, and, starting with one of the dots between thumb and forefinger of the right hand, fold the second crease to the left over on it, and do the same with each of the other dots. Place it, thus folded, in the funnel, moisten it, and fit to the side of the funnel. The filter will then have four short segments where there are three thicknesses and four where there is one thickness, but the latter are evenly distributed around its circumference, thus greatly aiding the passage of liquids through the paper and hastening both filtration and washing of the whole contents of the filter.

!SAMPLE PAGES FOR LABORATORY RECORDS!

!Page A!

Date

CALIBRATION OF BURETTE No.

BURETTE READINGS	DIFFERENCE	OBSERVED WEIGHTS	DIFFERENCE	CALCULATED CORRECTION
0.02		16.27		
10.12	10.10	26.35	10.08	-.02
20.09	9.97	36.26	9.91	-.06
30.16	10.07	46.34	10.08	+.01
40.19	10.03	56.31	9.97	-.06
50.00	9.81	66.17	9.86	+.05

These data to be obtained in duplicate for each burette.

!Page B!

Date

DETERMINATION OF COMPARATIVE STRENGTH HCl vs. NaOH

DETERMINATION	I	II
	Corrected	Corrected
Final Reading HCl	48.17 48.08	43.20 43.14
Initial Reading HCl	0.12 .12	.17 .17
	— —- — —-	— —- — —-
	47.96	42.97
	Corrected	Corrected
Final Reading HCl	46.36 46.29	40.51 40.37
Initial Reading HCl	1.75 1.75	.50 .50
	— —- — —-	— —- — —-
	44.54	39.87
log cc. NaOH	1.6468	1.6008
colog cc. HCl	8.3192	8.3668
	— — —	— — —
	9.9680 - 10	9.9676 - 10
1 cc. HCl	.9290 cc. NaOH	.9282 cc. NaOH
Mean	.9286	

Signed

Date

STANDARDIZATION OF HYDROCHLORIC ACID

Weight sample and tube	9.1793	8.1731
	8.1731	6.9187
	— — —	— — —
Weight sample	1.0062	1.2544
Final Reading HCl	39.97 39.83	49.90 49.77
Initial Reading HCl	.00 .00	.04 .04
	— —- — —-	— —- — —-
	39.83	49.73
Final Reading NaOH	.26 .26	.67 .67
Initial Reading NaOH	.12 .12	.36 .36
	—- —-	—- —-
	.14	.31
	.14	.31
Corrected cc. HCl	39.83 - — —- = 39.68	49.73 - — —- = 49.40
	.9286	.9286
log sample	0.0025	0.0983
colog cc	8.4014 - 10	8.3063 - 10
colog milli equivalent	1.2757	1.2757
	— — —	— — —
	9.6796 - 10	9.6803 - 10
Normal value HCl	.4782	.4789
Mean	.4786	

Signed

!Page D!
Date

DETERMINATION OF CHLORINE IN CHLORIDE, SAMPLE No.

Weight sample and tube	16.1721	15.9976
	15.9976	15.7117
	— — —-	— — —-
Weight sample	.1745	.2859
Weight crucible		
+ precipitate	14.4496	15.6915
Constant weights	14.4487	15.6915
	14.4485	
Weight crucible	14.2216	15.3196
Constant weight	14.2216	15.3194
Weight AgCl	.2269	.3721
log Cl	1.5496	1.5496
log weight AgCl	9.3558 - 10	9.5706 - 10
log 100	2.0000	2.0000
colog AgCl	7.8438 - 10	7.7438 - 10
colog sample	0.7583	0.5438
	— — —-	— — —-
	1.5075	1.5078
Cl in sample No.	32.18%	32.20%

Signed

STRENGTH OF REAGENTS

The concentrations given in this table are those suggested for use in the procedures described in the foregoing pages. It is obvious, however, that an exact adherence to these quantities is not essential.

	Approx. Grams per liter.	Approx. relation to normal solution	relation to molal solution
Ammonium oxalate, $(NH_4)_2C_2O_4 \cdot H_2O$	40	0.5N	0.25
Barium chloride, $BaCl_2 \cdot 2H_2O$	25	0.2N	0.1
Magnesium ammonium chloride (of $MgCl_2$)	71	1.5N	0.75
Mercuric chloride, $HgCl_2$	45	0.33N	0.66
Potassium hydroxide, KOH (sp. gr. 1.27)	480		
Potassium thiocyanate, KSCN	5	0.05N	0.55
Silver nitrate, $AgNO_3$	21	0.125N	0.125
Sodium hydroxide, NaOH	100	2.5N	2.5
Sodium carbonate. Na_2CO_3	159	3N	1.5
Sodium phosphate, $Na_2HPO_4 \cdot 12H_2O$	90	0.5N or 0.75N	0.25

Stannous chloride, $SnCl_2$, made by saturating hydrochloric acid (sp. gr. 1.2) with tin, diluting with an equal volume of water, and adding a slight excess of acid from time to time. A strip of metallic tin is kept in the bottle.

A solution of ammonium molybdate is best prepared as follows: Stir 100 grams of molybdic acid (MoO_3) into 400 cc. of cold, distilled water. Add 80 cc. of concentrated ammonium hydroxide (sp. gr. 0.90). Filter, and pour the filtrate slowly, with constant stirring, into a mixture of 400 cc. concentrated nitric acid (sp. gr. 1.42) and 600 cc. of water. Add to the mixture about 0.05 gram of microcosmic salt. Filter, after allowing the whole to stand for 24 hours.

The following data regarding the common acids and aqueous ammonia are based upon percentages given in the Standard Tables of the Manufacturing Chemists' Association of the United States

[!J.S.C.I.!, 24 (1905), 787-790]. All gravities are taken at 15.5°C. and compared with water at the same temperature.

Aqueous ammonia (sp. gr. 0.96) contains 9.91 per cent NH_3 by weight, and corresponds to a 5.6 N and 5.6 molal solution.

Aqueous ammonia (sp. gr. 0.90) contains 28.52 per cent NH_3 by weight, and corresponds to a 5.6 N and 5.6 molal solution.

Hydrochloric acid (sp. gr. 1.12) contains 23.81 per cent HCl by weight, and corresponds to a 7.3 N and 7.3 molal solution.

Hydrochloric acid (sp. gr. 1.20) contains 39.80 per cent HCl by weight, and corresponds to a 13.1 N and 13.1 molal solution.

Nitric acid (sp. gr. 1.20) contains 32.25 per cent HNO_3 by weight, and corresponds to a 6.1 N and 6.1 molal solution:

Nitric acid (sp. gr. 1.42) contains 69.96 per cent HNO_3 by weight, and corresponds to a 15.8 N and 15.8 molal solution.

Sulphuric acid (sp. gr. 1.8354) contains 93.19 per cent H_2SO_4 by weight, and corresponds to a 34.8 N or 17.4 molal solution.

Sulphuric acid (sp. gr. 1.18) contains 24.74 per cent H_2SO_4 by weight, and corresponds to a 5.9 N or 2.95 molal solution.

The term !normal! (N), as used above, has the same significance as in volumetric analyses. The molal solution is assumed to contain one molecular weight in grams in a liter of solution.

DENSITIES AND VOLUMES OF WATER AT TEMPERATURES FROM 15-30°C.

Temperature Density. Volume.
Centigrade.

Temperature	Density	Volume
4°	1.000000	1.000000
15°	0.999126	1.000874
16°	0.998970	1.001031
17°	0.998801	1.001200
18°	0.998622	1.001380
19°	0.998432	1.001571
20°	0.998230	1.001773
21°	0.998019	1.001985
22°	0.997797	1.002208
23°	0.997565	1.002441
24°	0.997323	1.002685
25°	0.997071	1.002938
26°	0.996810	1.003201
27°	0.996539	1.003473
28°		

0.996259 1.003755 29° 0.995971 1.004046 30°
0.995673 1.004346

Authority: Landolt, Börnstein, and Meyerhoffer's !Tabellen!, third edition.

CORRECTIONS FOR CHANGE OF TEMPERATURE OF STANDARD SOLUTIONS

The values below are average values computed from data relating to a considerable number of solutions. They are sufficiently accurate for use in chemical analyses, except in the comparatively few cases where the highest attainable accuracy is demanded in chemical investigations. The expansion coefficients should then be carefully determined for the solutions employed. For a compilation of the existing data, consult Landolt, Börnstein, and Meyerhoffer's !Tabellen!, third edition.

Corrections for 1 cc.
Concentration. of solution between
15° and 35°C.

Normal .00029
0.5 Normal .00025
0.1 Normal or more dilute solutions .00020

The volume of solution used should be multiplied by the values given, and that product multiplied by the number of degrees which the temperature of the solution varies from the standard temperature selected for the laboratory. The total correction thus found is subtracted from the observed burette reading if the temperature is higher than the standard, or added, if it is lower. Corrections are not usually necessary for variations of temperature of 2°C. or less.

INTERNATIONAL ATOMIC WEIGHTS

===
=========

| | |
| 1920 | | 1920
_____|_____|_____|_____
| | |

Aluminium Al | 27.1 | Molybdenum Mo | 96.0
Antimony Sb | 120.2 | Neodymium Nd | 144.3
Argon A | 39.9 | Neon Ne | 20.2
Arsenic As | 74.96 | Nickel Ni | 58.68
Barium Ba | 137.37 | Nitrogen N | 14.008
Bismuth Bi | 208.0 | Osmium Os | 190.9
Boron B | 11.0 | Oxygen O | 16.00
Bromine Br | 79.92 | Palladium Pd | 106.7
Cadmium Cd | 112.40 | Phosphorus P | 31.04
Caesium Cs | 132.81 | Platinum Pt | 195.2
Calcium Ca | 40.07 | Potassium K | 39.10
Carbon C | 12.005 | Praseodymium Pr | 140.9
Cerium Ce | 140.25 | Radium Ra | 226.0
Chlorine Cl | 35.46 | Rhodium Rh | 102.9
Chromium Cr | 52.0 | Rubidium Rb | 85.45
Cobalt Co | 58.97 | Ruthenium Ru | 101.7
Columbium Cb | 93.1 | Samarium Sm | 150.4
Copper Cu | 63.57 | Scandium Sc | 44.1
Dysprosium Dy | 162.5 | Selenium Se | 79.2
Erbium Er | 167.7 | Silicon Si | 28.3
Europium Eu | 152.0 | Silver Ag | 107.88
Fluorine Fl | 19.0 | Sodium Na | 23.00
Gadolinium Gd | 157.3 | Strontium Sr | 87.63
Gallium Ga | 69.9 | Sulphur S | 32.06
Germanium Ge | 72.5 | Tantalum Ta | 181.5
Glucinum Gl | 9.1 | Tellurium Te | 127.5
Gold Au | 197.2 | Terbium Tb | 159.2
Helium He | 4.00 | Thallium Tl | 204.0
Hydrogen H | 1.008 | Thorium Th | 232.4
Indium In | 114.8 | Thulium Tm | 168.5
Iodine I | 126.92 | Tin Sn | 118.7
Iridium Ir | 193.1 | Titanium Ti | 48.1
Iron Fe | 55.84 | Tungsten W | 184.0
Krypton Kr | 82.92 | Uranium U | 238.2
Lanthanum La | 139.0 | Vanadium V | 51.0

Lead Pb | 207.2 | Xenon Xe | 130.2
Lithium Li | 6.94 | Ytterbium Yb | 173.5
Lutecium Lu | 175.0 | Yttrium Y | 88.7
Magnesium Mg | 24.32 | Zinc Zn | 65.37
Manganese Mn | 54.93 | Zirconium Zr | 90.6
Mercury Hg | 200.6 | |
===

INDEX

Acidimetry
Acid solutions, normal
  standard
Acids, definition of
Acids, weak, action of other acids on
  action of salts on
Accuracy demanded
Alkalimetry
Alkali solutions, normal
  standard
Alumina, determination of in stibnite
Ammonium nitrate, acid
Analytical chemistry, subdivisions of
Antimony, determination of, in stibnite
Apatite, analysis of
Asbestos filters
Atomic weights, table of

   Balances, essential features of use and care of Barium sulphate, determination of sulphur in Bases, definition of Bichromate process for iron Bleaching powder, analysis of Brass, analysis of Burette, description of calibration of cleaning of reading of

Calcium, determination of, in limestone
Calibration, definition of
  of burettes
  of flasks
Carbon dioxide, determination of, in limestone
Chlorimetry
Chlorine, gravimetric determination of
Chrome iron ore, analysis of
Coin, determination of silver in
Colloidal solution of precipitates
Colorimetric analyses, definition of

249

Copper, determination of, in brass
  determination of in copper ores
Crucibles, use of
Crystalline precipitates

Densities of water
Deposition potentials
Desiccators
Direct methods
Dissociation, degree of

Economy of time
Electrolytic dissociation, theory of
Electrolytic separations, principles of
End-point, definition of
Equilibrium, chemical
Evaporation of liquids

Faraday's law
Feldspar, analysis of
Ferrous ammonium sulphate, analysis of
Filters, folding of
 how fitted
Filtrates, testing of
Filtration
Flasks, graduation of
Funnels
Fusions, removal of from crucibles

General directions for gravimetric analysis
  volumetric analysis
Gooch filter
Gravimetric analysis, definition of

Hydrochloric acid, standardization of
Hydrolysis

Ignition of precipitates Indicators, definition of for acidimetry preparation of Indirect methods Insoluble matter, determination of in limestone Integrity Iodimetry Ions, definition of Iron, gravimetric determination of volumetric determination of

Jones reductor

Lead, determination of in brass
Limestone, analysis of
Limonite, determination of iron in
Liquids, evaporation of
 transfer of
Litmus
Logarithms

Magnesium, determination of
Mass action, law of
Measuring instruments
Methyl orange
Moisture, determination of in limestone

Neutralization methods Normal solutions, acid and alkali oxidizing agents reducing agents Notebooks, sample pages of

Oxalic acid, determination of strength of
Oxidation processes
Oxidizing power of pyrolusite

Permanganate process for iron Phenolphthalein Phosphoric anhydride, determination of Pipette, calibration of description of Platinum crucibles, care of Precipitates, colloidal crystalline ignition of separation from filter washing of Precipitation Precipitation methods (volumetric) Problems Pyrolusite, oxidizing power of

Quantitative Analyses, subdivisions of

Reagents, strength of
Reducing solution, normal
Reductor, Jones
Reversible reactions

Silica, determination of, in limestone
  determination of, in silicates
  purification of
Silicic acid, dehydration of
Silver, determination of in coin
Soda ash, alkaline strength of
Sodium chloride, determination of chlorine in
Solubility product
Solution pressure
Solutions, normal
  standard
Standardization, definition of
Standard solutions, acidimetry and alkalimetry
  chlorimetry
  iodimetry
  oxidizing and reducing agents
  thiocyanate
Starch solutions
Stibnite, determination of antimony in
Stirring rods
Stoichiometry
Strength of reagents
Suction, use of
Sulphur, determination of in ferrous ammonium sulphate
  in barium sulphate

Temperature, corrections for
Testing of washings
Theory of electrolytic dissociation
Thiocyanate process for silver
Titration, definition of
Transfer of liquids

Volumetric analysis, definition of
  general directions

Wash-bottles
Washed filters
Washing of precipitates

Washings, testing of
Water, ionization of
 densities of
Weights, care of

Zimmermann-Reinhardt method for iron
Zinc, determination of, in brass